AVR

AVR

An Introductory Course

John Morton

Newnes

OXFORD AMSTERDAM BOSTON LONDON NEW YORK PARIS SAN DIEGO
SAN FRANCISCO SINGAPORE SYDNEY TOKYO

Newnes is an imprint of Elsevier
Linacre House, Jordan Hill, Oxford OX2 8DP, UK
30 Corporate Drive, Suite 400, Burlington, MA 01803, USA

First edition 2002
Reprinted 2005, 2007 (twice)

British Library Cataloguing in Publication Data
Morton, John, 1980–
  AVR: an introductory course
  1. Programmable controllers   2. Intergrated circuits – Design and construction
  3. Microelectronics
  I. Title
  626.8'9

Library of Congress Cataloging-in-Publication Data
A catalog record for this book is available from the Library of Congress

ISBN: 978-0-7506-5635-1

For information on all Newnes publications
visit our website at www.newnespress.com

Printed and bound in *Great Britain*

07 08 09 10 10 9 8 7 6 5 4

Contents

For Tara

Acknowledgements

Robert Czarnek first introduced me to AVRs, and I quickly recognized their strengths over other microcontrollers. The only relative weakness that I saw in them was a lack of fame on the scale of the Microchip's PIC, for example. I knew it was only a matter of time before this would be steadily overcome and this book is a guide for those with little or no microcontroller background to start using AVRs.

I would like to take this opportunity to thank those who have assisted me with what you see before you. Atmel UK kindly donated a sample of their equipment, though I assure you I remain impartial and objective! A big thanks must go to Matt Webb for his efficient and meticulous proofreading, which often consisted of writing 'What's this?' all over the page. He really had much better things to do, for example passing his finals, but still managed to tear himself away to comb through my pages. I would also like to thank Richard George for his suggestions of example projects and general ideas. Thanks to Matt Harrison for his help with the illustrations – he is off to further this calling at the Royal College of Art. Finally, I must thank Max Horsey for his great generosity, assistance and advice, and also the Electronics Department at Radley College, Abingdon, for the continuing use of their excellent lab.

John Morton

Preface

Congratulations! By reading this you're showing an interest in one of the most capable and versatile 8-bit microcontrollers on the market, the AVR. Continue reading this book to learn about the entire AVR family, and how they can help simplify the design of your electronics projects as well as allow you to create more sophisticated products.

Like all microcontrollers, AVRs allow tailor-made solutions which remain at the same time completely flexible. However, AVRs are efficient, fast, and easy to use microcontrollers, making them an ideal choice for designers. In this book I begin from the most basic principles of microcontroller programming, such as binary and hexadecimal, and cover the principal steps in developing a program. Each AVR topic is introduced alongside one of twenty worked examples, which include a pedestrian-crossing simulator, melody generator, frequency counters and a computer-controlled robot.

To begin with, the programs are largely developed for you. However, as you progress through each chapter, more and more of the programs will be written by you in the form of the exercises, which appear throughout the book with answers given at the end of the book. The appendices summarize the key properties of the most popular AVRs allowing quick reference without having to plough through piles of datasheets.

In short this book offers a hands on approach to learning to program AVRs, and will provide a useful source of information for AVR programmers.

John Morton

1
Introduction

An AVR is a type of microcontroller, and not just any microcontroller – AVRs are some of the fastest around. I like to think of a microcontroller as a useless lump of silicon with amazing potential. It will do nothing without but almost anything with the program that you write. Under your guidance, a potentially large conventional circuit can be squeezed into one program and thus into one chip. Microcontrollers bridge the gap between hardware and software – they run programs, just like your computer, yet they are small, discrete devices that can interact with components in a circuit. Over the years they have become an indispensable part of the toolbox of electrical engineers and enthusiasts as they are perfect for experimenting, small batch productions, and projects where a certain flexibility of operation is required.

Figure 1.1 shows the steps in developing an AVR program.

1. The blank AVR does nothing

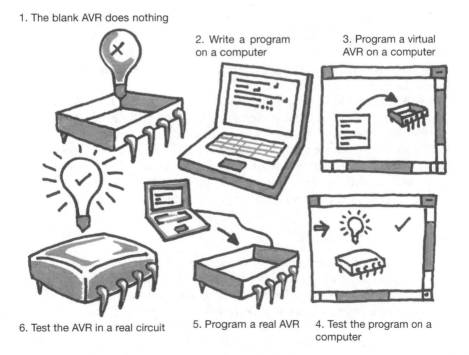

2. Write a program on a computer

3. Program a virtual AVR on a computer

6. Test the AVR in a real circuit

5. Program a real AVR

4. Test the program on a computer

Figure 1.1

The AVR family covers a huge range of different devices, from Tiny 8-pin devices to the Mega 40-pin chips. One of the fantastic things about this is that you can write a program with one type of AVR in mind, and then change your mind and put the program in a different chip with only minimal changes. Furthermore, when you learn how to use one AVR, you are really learning how to use them all. Each has its own peculiarities – their own special features – but underneath they have a common heart.

Fundamentally, AVR programming is all to do with pushing around numbers. The trick to programming, therefore, lies in making the chip perform the designated task by the simple movement and processing of numbers. There is a specific set of tasks you are allowed to perform on the numbers – these are called *instructions*. The program uses simple, general instructions, and also more complicated ones which do more specific jobs. The chip will step through these instructions one by one, performing millions every second (this depends on the frequency of the oscillator it is connected to) and in this way perform its job. The numbers in the AVR can be:

1. **Received** from inputs (e.g. using an input 'port')
2. **Stored** in special compartments inside the chip
3. **Processed** (e.g. added, subtracted, ANDed, multiplied etc.)
4. **Sent out** through outputs (e.g. using an output 'port')

This is essentially all there is to programming ('great' you may be thinking). Fortunately there are certain other useful functions that the AVR provides us with such as on-board timers, serial interfaces, analogue comparators, and a host of 'flags' which indicate whether or not something particular has happened, which make life a lot easier.

We will begin by looking at some basic concepts behind microcontrollers, and quickly begin some example projects on the AT90S1200 (which we will call 1200 for short) and Tiny AVRs. Then intermediate operations will be introduced, with the assistance of more advanced chips (such as the AT90S2313). Finally, some of the more advanced features will be discussed, with a final project based around the 2313. Most of the projects can be easily adapted for any type of AVR, so there is no need for you to go out and buy all the models.

Short bit for PIC users

A large number of readers will be familiar with the popular PIC microcontroller. For this reason I'll mention briefly how AVRs can offer an improvement to PICs. For those of you who don't know what PICs are, don't worry too much if you don't understand all this, it will all make sense later on!

Basically, the AVRs are based on a more advanced underlying architecture, and can execute an instruction every clock cycle (as opposed to PICs which execute one every four clock cycles). So for the same oscillator frequency, the AVRs will run four times as fast. Furthermore they also offer 32 working registers (compared with the one that PICs have), and about three times as many instructions, so programs will almost always be shorter. It is worth noting, however, that although the datasheets boast 90–120 instructions, there is considerable repetition and redundancy, and so in my view there are more like 50 distinct instructions.

Furthermore, what are known as special function registers on PICs (and known as input/output registers on the AVR) can be directly accessed with PICs (e.g. you can write directly to the ports), and this cannot be done to the same extent with AVRs. However, these are minor quibbles, and AVR programs will be more efficient on the whole. All AVRs have flash program memory (so can be rewritten repeatedly), and finally, as the different PICs have been developed over a period of many years there are some annoying compatibility issues between some models which the AVRs have managed to avoid so far.

Number systems

It is worth introducing at this stage the different numbering systems which are involved in AVR programming: binary, decimal and hexadecimal. A binary number is a *base 2* number (i.e. there are only two types of digit (0 and 1)) as opposed to decimal – *base 10* – with 10 different digits (0 to 9). Likewise hexadecimal represents *base 16* so it has 16 different digits (0, 1, 2, 3, 4, 5, 6, 7, 8, 9, A, B, C, D, E and F). The table below shows how to count using the different systems:

binary (8 digit)	decimal (3 digit)	hexadecimal (2 digit)
00000000	000	00
00000001	001	01
00000010	002	02
00000011	003	03
00000100	004	04
00000101	005	05
00000110	006	06
00000111	007	07
00001000	008	08
00001001	009	09
00001010	010	0A
00001011	011	0B
00001100	012	0C
00001101	013	0D
00001110	014	0E
00001111	015	0F
00010000	016	10
00010001	017	11
etc.		

The binary digit (or *bit*) furthest to the right is known as the least significant bit or *lsb* and also as *bit 0* (the reason the numbering starts from 0 and not from 1 will soon become clear). Bit 0 shows the number of 'ones' in the number. One equals 2^0. The bit to its left (*bit 1*) represents the number of 'twos', the next one (*bit 2*) shows the number of 'fours' and so on. Notice how two $= 2^1$ and four $= 2^2$, so the bit number corresponds to the power of two which that bit represents, but note that the numbering goes from right to left (this is very often forgotten!). A sequence of 8 bits is known as a byte. The highest number bit in a binary word (e.g. bit 7 in the case of a byte) is known as the most significant bit (*msb*).

So to work out a decimal number in binary you could look for the largest power of 2 that is smaller than that number and work your way down.

Example 1.1   Work out the binary equivalent of the decimal number 83.
Largest power of two less than $83 = 64 = 2^6$. Bit 6 = **1**

This leaves $83 - 64 = 19$	32 is greater than 19 so bit 5 = **0**,
	16 is less than 19 so bit 4 = **1**,
This leaves $19 - 16 = 3$	8 is greater than 3 so bit 3 = **0**,
	4 is greater than 3 so bit 2 = **0**,
	2 is less than 3 so bit 1 = **1**,
This leaves $3 - 2 = 1$	1 equals 1 so bit 0 = **1**.

So **1010011** is the binary equivalent.

There is, however, an alternative (and more subtle) method which you may find easier. Take the decimal number you want to convert and divide it by two. If there is a remainder of one (i.e. it was an odd number), write down a one. Then divide the result and do the same writing the remainder to the *left* of the previous value, until you end up dividing one by two, leaving a one.

Example 1.2   Work out the binary equivalent of the decimal number 83.

Divide 83 by two.	Leaves 41, remainder **1**
Divide 41 by two.	Leaves 20, remainder **1**
Divide 20 by two.	Leaves 10, remainder **0**
Divide 10 by two.	Leaves 5, remainder **0**
Divide 5 by two.	Leaves 2, remainder **1**
Divide 2 by two.	Leaves 1, remainder **0**
Divide 1 by two.	Leaves 0, remainder **1**

So **1010011** is the binary equivalent.

EXERCISE 1.1   Find the binary equivalent of the decimal number 199.

EXERCISE 1.2   Find the binary equivalent of the decimal number 170.

Likewise, bit 0 of a hexadecimal is the number of ones ($16^0 = 1$) and bit 1 is the number of 16s ($16^1 = 16$) etc. To convert decimal to hexadecimal (it is often abbreviated to just 'hex') look at how many 16s there are in the number, and how many ones.

Example 1.3   Convert the decimal number 59 into hexadecimal. There are 3 16s in 59, leaving $59 - 48 = 11$. So bit 1 is 3. 11 is B in hexadecimal, so bit 0 is B. The number is therefore **3B**.

EXERCISE 1.3   Find the hexadecimal equivalent of 199.

EXERCISE 1.4   Find the hexadecimal equivalent of 170.

One of the useful things about hexadecimal, which you may have picked up from Exercise 1.4, is that it translates easily with binary. If you break up a binary number into 4-bit groups (called *nibbles*, i.e. small bytes), these little groups can individually be translated into 1 hex digit.

Example 1.4   Convert 01101001 into hex. Split the number into nibbles: 0110 and 1001. It is easy to see 0110 translates as $4 + 2 = 6$ and 1001 is $8 + 1 = 9$. So the 8-bit number is **69** in hexadecimal. As you can see, this is much more straightforward than with decimal, which is why hexadecimal is more commonly used.

EXERCISE 1.5   Convert 11100111 into a hexadecimal number.

Adding in binary

Binary addition behaves in exactly the same way as decimal addition. Examine each pair of bits.

$$0 + 0 = 0 \qquad \text{no carry}$$
$$1 + 0 = 1 \qquad \text{no carry}$$
$$1 + 1 = 0 \qquad \text{carry 1}$$
$$1 + 0 + 0 = 1 \qquad \text{no carry}$$
$$1 + 1 + 0 = 0 \qquad \text{carry 1}$$
$$1 + 1 + 1 = 1 \qquad \text{carry 1}$$

Example 1.5 $4 + 7 = 11$

```
          1
          0100
   +      0111
          1011  = 11 in decimal
```

EXERCISE 1.6   Find the result of 01011010 + 00001111 using binary addition.

Negative numbers

We have seen how positive decimal numbers translate into binary, but how do we translate negative numbers? We have to sacrifice a bit towards giving the number a sign, so for a 4-bit signed number, the range of values might be −7 to +8. There are various representations for negative numbers, including *two's complement*. With this method, to make a positive number onto its negative equivalent, you invert all the bits and then add one:

Example 1.6   0111 = 7
                 Invert all bits: 1000
                 Add one: 1001
                 1001 = −7

Example 1.7   1000 = 8
                 Invert: 0111
                 Add one: 1000
                 1000 = −8 = +8   *FAIL!*

As you can see in Example 1.7, we cannot use −8 because it is indistinguishable from +8. This asymmetry is recognized as an unfortunate consequence of the two's complement method, but it has been accepted as the best given the shortcomings of other methods of signing binary numbers. Let's test these negative numbers by looking at −2 + 7:

Example 1.8   2 = 0010 therefore −2 = 1110

                 1110 = −2
                 + 0111 = 7
                 0101 = 5   Which is what we would expect!

EXERCISE 1.7   Find the 8-bit two's complement representation of −40, and show that −40 + 50 gives the expected result.

A result of this notation is that we can simply test the most significant bit (msb) to see whether a number is positive or negative. A 1 in the msb indicates a negative number, and a 0 indicates positive. However, when dealing with the result of addition and subtraction with large positive or negative numbers, this can be misleading.

Example 1.9   69 + 120 = ...

$$
\begin{array}{r}
1\\
01000101 = +69\\
+\quad 01111000 = +120\\
\hline
10111101 = +189 \text{ or } -67
\end{array}
$$

In other words, in the two's complement notation, we could interpret the result as having the msb 1 and therefore negative. There is therefore a test for 'two's complement overflow' which we can use to determine the *real* sign of the result. The 'two's complement overflow' occurs when:

- both the msb's of the numbers being added are 0 and the msb of the result is 1
- both the msb's of the numbers being added are 1 and the msb of the result is 0

The *real* sign is therefore given by a combination of the 'two's complement overflow' result, and the state of the msb of the result:

Two's complement overflow?	MSB of result	Sign
No	0	Positive
No	1	Negative
Yes	0	Negative
Yes	1	Positive

As you can see from Example 1.10, there is a two's complement overflow, and the msb of the result is 1, and so the sign of the answer is positive (+189) as we would expect. You will be relieved to hear that much of this is handled automatically by the AVR.

The *one's complement* is simply the result of inverting all the bits in a number.

An 8-bit RISC Flash microcontroller?

We call the AVR an *8-bit microcontroller*. This means it deals with numbers 8 bits long. The binary number 11111111 is the largest 8-bit number and equals 255 in decimal and FF in hex (work it out!). With AVR programming, different notations are used to specify different numbering systems (the decimal number 11111111 is very different from the binary number 11111111)! A binary number is shown like this: 0b00101000 (i.e. **0b**...). Decimal is the default system, and the hexadecimal numbers are written with a **0x**, or with a dollar sign, like this: 0x3A or $3A. Therefore:

0b00101011   is equivalent to   43   which is equivalent to   0x2B

When dealing with the inputs and outputs of an AVR, binary is always used, with each input or output pin corresponding to a particular bit. A **1** corresponds to what is known as *logic 1*, meaning the pin of the AVR is at the supply voltage (e.g. +5 V). A 0 shows that the pin is at *logic 0*, or 0 V. When used as inputs, the boundary between reading a logic 0 and a logic 1 is half of the supply voltage (e.g. +2.5 V).

You will also hear the AVR called a *RISC microcontroller*. This means it is a **R**educed **I**nstruction **S**et **C**omputer, i.e. has relatively few instructions. This makes life slightly harder for the programmer (you or me), but the chip itself is more simple and efficient.

The AVR is sometimes called a *Flash microcontroller*. This refers to the fact that the program you write for it is stored in *Flash memory* – memory which can be written to again and again. Therefore you can keep reprogramming the same AVR chip – for hobbyists this means one chip can go a long way.

Initial steps

The process of developing a program consists of five basic steps:

1. **Select** a particular AVR chip, and construct a program **flowchart**
2. **Write** program (using Notepad, AVR Studio, or some other suitable development software)
3. **Assemble** program (changes what you've written into something an AVR will understand)
4. **Simulate** or **Emulate** the program to see whether or not it works
5. **Program** the AVR. This feeds what you've written into the actual AVR

Let's look at some of these in more detail.

Choosing your model

As there are so many different AVRs to choose from, it is important you think carefully about which one is right for your application. The name of the AVR can tell you some information about what it has, e.g.:

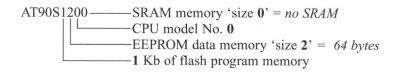

AT90S1200 ——— SRAM memory 'size **0**' = *no SRAM*
              └——— CPU model No. **0**
              └——— EEPROM data memory 'size **2**' = *64 bytes*
              └——— **1** Kb of flash program memory

Memory sizes:

0	1	2	3	4	5	6	7	8	9	A	B
0	32	64	128	256	512	1K	2K	4K	8K	16K	32K
	bytes	bytes	bytes	bytes	bytes						

The meaning of these terms may not be familiar, but they will be covered shortly. The Tiny and Mega family have slightly different systems. You can get a decent overview of some of the AVRs and their properties by checking out Appendix A.

EXERCISE 1.8   Deduce the memory properties of the AT90S8515.

One of the most important features of the AVR, which unfortunately is not encoded in the model name, is the number of input and output pins. The 1200 has 15 input/output pins (i.e. they have 15 pins which can be used as inputs *or* outputs), and the 8515 has up to 32!

Example 1.10   The brief is to design a device to count the number of times a push button is pressed and display the value on a single seven segment display – when the value reaches nine it resets.

1. The seven segment display requires **seven** outputs
2. The push button requires **one** input

This project would therefore need a total of eight input/output pins. In this case a 1200 would be used as it is one of the simplest models and has enough pins.
  A useful trick when dealing with a large number of inputs and outputs is called **strobing**. It is especially handy when using more than one seven segment display, or when having to test many buttons. An example demonstrates it best.

Example 1.11   The brief is to design a counter which will add a number between 1 and 9 to the current two-digit value. There are therefore nine push buttons and two seven segment displays.
  It would first appear that quite a few inputs and outputs are necessary:

1. The two seven segment displays require seven outputs each, thus a total of **14**
2. The push buttons require one input each. Creating a total of **nine**

The overall total is therefore 23 input/output pins, which would require a large AVR such as the 8515 (which has 32 I/O pins); however, it would be unnecessary to use such a large one as this value can be cut significantly.
  By strobing the buttons, they can all be read using only six pins, and the two

seven segment displays can be controlled by only nine. This creates a total of 15 input/output (or I/O) pins, which would just fit on the 1200. Figure 1.2 shows how it is done.

By making the pin labelled PB0 logic 1 (+5 V) and PB1, PB2 logic 0 (0 V), switches 1, 4 and 7 are enabled. They can then be tested individually by examining pins PB3 to PB5. Thus by making PB0 to PB2 logic 1 one by one, all the buttons can be examined individually. In order to work out how many I/O pins you will need for an array of X buttons, find the pair of factors of X which have the smallest sum (e.g. for 24, 6 and 4 are the factors with the smallest sum, hence 6 + 4 = 10 I/O pins will be needed). It is better to make the smaller of the two numbers (if indeed they are not the same) the number of outputs, and the larger the number of inputs. This way the program takes less time to scroll through all of the rows of buttons.

Strobing seven segment displays basically involves displaying a number on one display for a short while, and then turning that display off while you display another number on another display. PD0 to PD6 contain the seven segment code for both displays, and by making PB6 or PB7 logic 1, you can turn the individual displays on. So the displays are in fact flashing on and off at high speed, giving the impression that they are constantly on. The programming requirements of such an arrangement will be examined at a later stage.

EXERCISE 1.9 With the help of Appendix A, work out which model AVR you would use for a four-digit calculator with buttons for digits 0–9 and five operations: +, −, ×, ÷ and = .

Flowchart

After you have worked out how many I/O pins you will need, and thus selected a particular AVR, the next step is to create a program flowchart. This basically forms the backbone of a program, and it is much easier to write a program from a flowchart than from scratch.

A flowchart should show the fundamental steps that the AVR must perform and a clear program structure. Picture your program as a hedge maze. The flowchart is a rough map showing key regions of the maze. When planning your flowchart you must note that the maze cannot lead off a cliff (i.e. the program cannot simply end), or the AVR will run over the edge and crash. Instead the AVR is doomed to navigate the maze indefinitely (although you can send it to sleep!). A simple example of a flowchart is shown in Figure 1.3.

Example 1.12 The flowchart for a program to turn an LED on *if* a button is being pressed.

(The Set-up box represents some steps which must be taken as part of the start of every program, in order to set up various functions – this will be examined

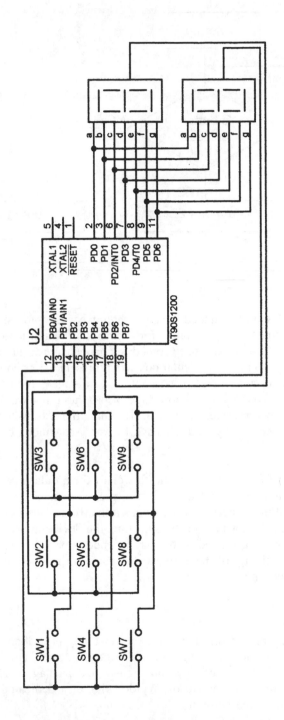

Figure 1.2

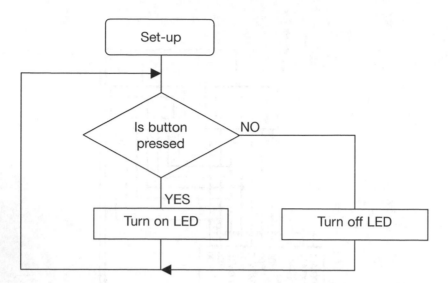

Figure 1.3

later.) Rectangles with rounded corners should be used for start and finish boxes, and diamond-shaped ones for decisions. Conditional jumps (the diamond shaped boxes) indicate '*if* something happens, *then* jump somewhere'.

The amount of code any particular box will represent varies considerably, and is really not important. The idea is to get the key stages, and come up with a diagram that someone with no knowledge of programming would understand. You will find it much easier to write a program from a flowchart, as you can tackle each box separately, and not have to worry so much about the overall structure.

EXERCISE 1.10   *Challenge!* Draw the flowchart for an alarm with three push buttons. Once the device is triggered by a pressure sensor, the three buttons must be pressed in the correct order, and within 10 seconds, or else the alarm will go off. If the buttons are pressed in time, the device returns to the state it was in before being triggered. If the wrong code is pressed the alarm is triggered. (The complexity of the answers will vary, but to give you an idea, my answer has 13 boxes.)

Writing

Once you have finished the flowchart, the next step is to load up a program template (such as the one suggested on page 19), and begin writing your program into it. This can be done on a basic text package such as Notepad (the one that comes with Microsoft Windows®), or a dedicated development environment such as AVR Studio.

Assembling

When you have finished writing your program, it needs to be *assembled* before it can be transferred onto a chip. This converts the program you've written into a series of numbers which can be fed into the Flash Program Memory of the AVR. This series of numbers is called the *hex code* or *hex file* – a hex file will have **.hex** after its name. The assembler will examine your program line by line and try to convert each line into the corresponding hex code. If, however, it fails to recognize something in one of the lines of your code, it will register an **error** for that line. An error is something which the assembler thinks is *definitely* wrong – i.e. it can't understand it. It may also register a **warning** – something which is probably wrong, i.e. definitely unusual but not necessarily wrong. All this should be made much more clear when we actually assemble our first program.

Registers

One of the most important aspects to programming with AVRs and microcontrollers in general are the *registers*. I like to think of the AVR as having a large filing cabinet with many drawers, each containing an 8-bit number (a byte). These drawers are registers – more specifically we call these the *I/O registers*. In addition to these I/O registers, we have 32 'working' registers – these are different because they are not part of the filing cabinet. Think of the working registers as the filing assistants, and yourself as the boss. If you want something put in the filing cabinet, you give it to the filing assistant, and then tell them to put it in the cabinet. In the same way, the program writer cannot move a number directly into an I/O register. Instead you must move the number into a working register, and then copy the working register to the I/O register. You can also ask your filing assistants to do arithmetic etc. on the numbers they hold – i.e. you can add numbers between working registers. Figure 1.4 shows the registers on the 1200.

As you can see, each register is assigned a number. The working registers are assigned numbers R0, R1, . . ., R31. Notice, however, that R30 and R31 are slightly different. They represent a double register called Z – an extra long register that can hold a 16-bit number (called a *word*). These are two filing assistants that can be tied together. They can be referred to independently – ZL and ZH – but can be fundamentally linked in that ZL (Z Lower) holds bits 0–7 of the 16-bit number, and ZH (Z Higher) holds bits 8–15.

Example 1.13

ZH	ZL				ZH	ZL
00000000	11111111	$\rightarrow$	add one to ZL	$\rightarrow$	00000001	00000000

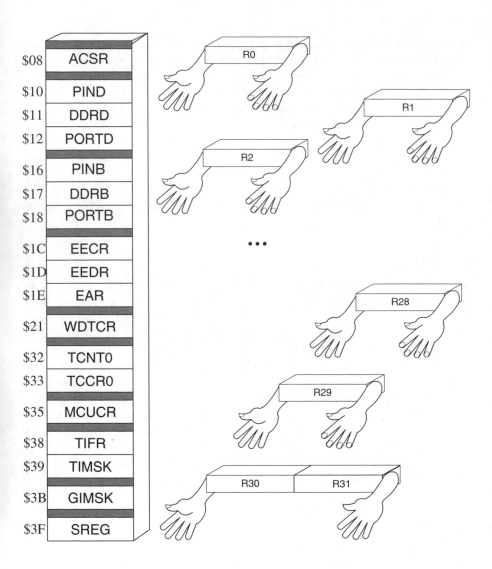

Figure 1.4

Example 1.14

ZH	ZL			ZH	ZL
11111111	11111111	→	add one to ZL →	00000000	00000000

Note that this linking only occurs with certain instructions. Assume that an instruction *doesn't* have the linking property unless explicitly stated.

You will find it easier to give your working registers names (for the same reason you don't call your filing assistants by their staff numbers), and you will be able to do this. It is sensible to give them a name according to the meaning of the number they are holding. For example, if you were to use register R5 to store the number of minutes that have passed, you might want to call it something like **Minutes**. You will be shown how to give names to your registers shortly, when we look at the program template. We will also see later that the working registers numbers R16–R31 are slightly more powerful than the others.

The I/O registers are also assigned numbers (0–63 in decimal, or $0–$3F in hexadecimal). Each of these performs some specific function (e.g. count the passage of time, or control serial communications etc.) and we will go through the function of each one in due course. I will, however, highlight the functions of PORTB, PORTD, PINB and PIND. These I/O registers represent the ports – the AVR's main link with the outside world. If you're wondering what happened to Ports A and C, it's not really very important. All four (A, B, C and D) appear on larger types of AVR (e.g. 8515); smaller AVRs (e.g. 1200) have only two. These two correspond to the two on larger AVRs that are called B and D, hence their names.

Figure 1.5 shows the pin layout of the 1200. Notice the pins labelled PB0, PB1, . . ., PB7. These are the Port B pins. Pins PD0–PD6 are the Port D pins. They can be read as inputs, or controlled as outputs. If behaving as an input, reading the binary number in PINB or PIND tells us the states of the pin, with PB0 corresponding to bit 0 in PINB etc. If the pin is high, the corresponding bit is 1, and vice versa. Note that Port D doesn't have the full 8 bits.

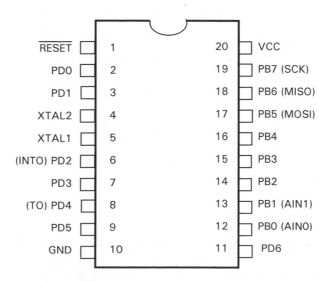

Figure 1.5

Example 1.15 All of PB0–PB7 are inputs. They are connected to push buttons which are in turn connected to the +5 V supply rail. When all the buttons are pressed, the number in PINB is 0b11111111 or 255 in decimal. When all buttons except PB7 are pressed, the number in PINB is 0b01111111 or 127 in decimal.

In a similar way, if the pin is an output its state is controlled by the corresponding bit in the PORTx register. The pins can sink or source 20 mA, and so are capable of driving LEDs directly.

Example 1.16 All of PB0–PB7 are outputs connected to LEDs. The other legs of the LEDs are connected to ground (via resistors). To turn on all of the LEDs, the number 0b11111111 is moved into PORTB. To turn off the middle two LEDs, the number 0b11100111 is moved into PORTB.

EXERCISE 1.11 Consider the example given above where all of PB0–PB7 are connected to LEDs. We wish to create a chase of the eight LEDs (as shown in Figure 1.6), and plan to move a series of numbers into PORTB one after the other to create this effect. What will these numbers be (in binary, decimal and hexadecimal)?

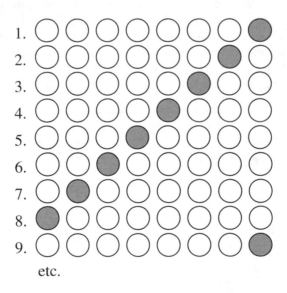

etc.

Figure 1.6

EXERCISE 1.12 PD0, PD1 and PD2 are connected to push buttons which are in turn connected to the +5 V supply rail. These push buttons are used in a controller for a quiz show. What numbers in PIND indicate that more than one button is being pressed at one time (in binary, decimal and hexadecimal)?

Instructions

We will now begin looking at some instructions. These are summarized in Appendix C at the back of the book. AVRs generally have about a hundred different instructions supported on them. This may sound quite daunting at first, but you will be relieved to hear that there is a fair amount of repetition. In fact there are only really about 40 that you *really* need to remember, and many are quite easy to remember with familiar sounding names like **add** or **jmp**. Fortunately, there a few general rules to help you decipher an unknown instruction. First, whenever you come across the letter **i** in an instruction, it will often stand for *immediate*, i.e. the number which *immediately* follows the instruction or **I/O** register. A **b** will often stand for **b**it or **b**ranch (i.e. jump to a part of the program). Let's take a look at the format of an instruction line.

Example 1.17

(Label:)   sbi   portb, 0   ; turns on LED

The optional first part of the line is the label. This allows another part of the program to jump to this line. Note that a label cannot start with a number, and should not be given the same name as an instruction, or a file register (as this will confuse the AVR greatly!). The label is always immediately followed by a colon (this is easy to leave off and can be a common source of errors if you aren't careful). Note that the label doesn't actually have to be on the same line as the instruction it's labelling. For example, the following is just as valid:

Label:
              sbi   portb, 0   ; turns on LED

After the label comes the actual instruction: **sbi**, i.e. what you are *doing*, and then comes what you are doing it *to*: **portb, 0** (these are called the operands). Lastly, and just as important, is a semicolon followed by a comment on what the line is actually doing in your own words. It is worth noting that you can write whatever you want in an AVR program as long as it comes *after* a semicolon. Otherwise the assembler will try to translate what you've written (e.g. 'turns on LED') and obviously fail and register an ERROR. As the assembler scans the program line by line, it skips to the next line when it encounters a semicolon.

I must stress how important it is to *explain* every line you write, as shown above. There are a number of reasons for this. First, what you've written may make sense to you as you write it, but after a few coffee breaks, or a week later, or a month later, you'll be looking at the line and wondering what on earth you were intending to do. Second, you may well be showing your program to other people for advice. I am sent programs that, with alarming regularity, contain very few or in some cases no comments at all. There is not much one can do in this situation, as it is almost impossible to deduce the intended operation of the program by looking at the bare code. Writing good comments is not necessarily easy – they should be very clear, but not too long. It is particularly worth avoiding falling into the habit of just copying out the meaning of the line.

Example 1.18

sbi   PortB, 0   ; sets bit 0 of register PortB

A comment like the one above means very little at all, as it doesn't tell you *why* you're setting bit 0 of register PortB, which after all is what the comment is really about. If you want to get an overview of all the instructions offered, have a good look at Appendix C and you can get a feel of how the different instructions are arranged. They will be introduced one by one through the example projects which follow.

Program template

Most programs will have a certain overall structure, and there are certain common elements needed for all programs to work. To make life easier, therefore, we can put together a program template, save it, and then load it every time we want to start writing a program. A template that I like to use is shown in Figure 1.7.

The box made up of asterisks at the top of the template is the program header (the asterisks are there purely for decorative purposes). Filling these in makes it easier to find out what the program is without having to scroll down and read the code and it helps you ensure that you are working on the most up-to-date version of your program. Note that the contents of the box have no bearing on the actual functioning of your program, as all the lines are preceded by semi-colons. The 'clock frequency:' line refers to the frequency of the oscillator (e.g. crystal) that you have connected to the chip. The AVR needs a steady signal to tell it when to move on to the next instruction, and so executes an instruction for every oscillation (or *clock cycle*). Therefore, if you have connected a 4 MHz crystal to the chip, it should execute about 4 million instructions per second. Note that I say *about* 4 million, because some instructions (typically the ones which involve jumping around in the program) take *two* clock cycles. 'for AVR:' refers to which particular AVR the program is written for. You will also need to specify this further down.

Now we get to the lines which actually do something. **.device** is a *directive* (an instruction to the assembler) which tells the assembler which device you are using. For example, if you were writing this for the 1200 chip, the complete line would be:

```
;****************************************
; written by:                        *
; date:                              *
; version:                           *
; file saved as:                     *
; for AVR:                           *
; clock frequency:                   *
;****************************************

; Program Function:_____
;  _____

.device     xxxxxxxx
.nolist
.include        "C:\Program Files\Atmel\AVR Studio\Appnotes\xxxxxx.inc"
.list

;==========
; Declarations:

.def        temp        =r16

;==============
; Start of Program

        rjmp   Init                    ; first line executed

;==========
Init:   ldi             temp, 0bxxxxxxxx    ; Sets up inputs and outputs on PortB
        out             DDRB, temp          ;
        ldi             temp, 0bxxxxxxxx    ; Sets up inputs and outputs on PortD
        out             DDRD, temp          ;

        ldi             temp, 0bxxxxxxxx    ; Sets pulls ups for inputs of PortB
        out             PortB, temp         ;    and the initial states for the outputs
        ldi             temp, 0bxxxxxxxx    ; Sets pulls ups for inputs of PortD
        out             PortD, temp         ;    and the initial states for the outputs

;====================
; Main body of program:
Start:
        <write your program here>
        rjmp   Start                    ; loops back to Start
```

Figure 1.7

.device   at90s1200

Another important directive is **.include**, which enables the assembler to load what is known as a *look-up file*. This is like a translator dictionary for the assembler. The assembler will understand most of the terms you write, but it may need to *look up* the translations of others. For example, all the names of the input/output registers and their addresses are stored in the look-up file, so instead of referring to $3F, you can refer to SREG. When you install the assembler on your computer, it should come with these files and put them in a directory. I have included the path that appears on my own computer but yours may well be different. Again, if the 1200 was being used, the complete line would be:

.include   "C:\Program Files\Atmel\AVR Studio\Appnotes\1200def.inc"

Finally I'll say a little about **.nolist** and **.list**. As the assembler reads your code, it can produce what is known as a *list file*, which includes a copy of your program complete with the assembler's comments on it. By and large, you do not want this list file also to include the lengthy look-up file. You therefore write **.nolist** before the **.include** directive, which tells the assembler to stop copying things to the list file, and then you write **.list** after the **.include** line to tell the assembler to resume copying things to the list file. In summary, therefore, the .nolist and .list lines don't actually change the working of the program, but they will make your list file tidier. We will see more about list files when we begin our first program.

After the general headings, there is a space to specify some *declarations*. These are your own additions to the assembler's translator dictionary – your opportunities to give more useful names to the registers you will be using. For example, I always use a working register called **temp** for menial tasks, and I've assigned this name to R16. You can define the names of the working registers using the **.def** directive, as shown in the template. Another type of declaration that can be used to generally give a numerical value to a word is **.equ**. This can be used to give your own names to I/O registers. For example, I might have connected a seven segment display to all of Port B, and decided that I wish to be able to write DisplayPort when referring to PortB. PortB is I/O register number 0x18, so I might write DisplayPort in the program and the assembler will interpret it as PortB:

.equ   DisplayPort = PortB   or
.equ   DisplayPort = 0x18

Another example of where this might be useful is where a particular number is used at different points in the program, and you might be experimenting and changing this number. You could use the **.equ** directive to give a name to this number, and simply refer to the name in the rest of the program. When you then go to change the number, you need only change the value in the **.equ** line, and not in all the instances of the use of the number all over the program. For the moment, however, we will not be using the **.equ** directive.

After the declarations, we have the first line executed by the chip on power-up or reset. In this line I suggest jumping to a section called **Init** which sets up all the initial settings of the AVR. This uses the **rjmp** instruction:

   rjmp   Init   ;

This stands for **r**elative **jump**. In other words it makes the chip jump to a section of the program which you have labelled **Init**. The reason why it is a *relative* jump is in the way the assembler interprets the instruction, and so is not really important to understand. Say, for example, that the **Init** section itself was 40 instructions further on from the **rjmp   Init** line, the assembler would inter-pret the line as saying 'jump forward 40 instructions' – i.e. a jump *relative* to the original instruction. Basically it is far easier to think of it as simply jumping *to* **Init**.

The first part of the **Init** section sets which pins are going to act as inputs, and which as outputs. This is done using the Data Direction I/O registers: DDRB and DDRD. Each bit in these registers corresponds to a pin on the chip. For example, bit 4 of DDRB corresponds to pin PB4, and bit 2 of DDRD corre-sponds to pin PD2. Now, setting the relative DDRx bit *high* makes the pin an *output*, and making the bit *low* makes the pin an *input*.

If we configure a pin as an input, we then have the option of selecting whether the input has a built-in pull-up resistor or not. This may save us the trouble of having to include an external resistor. In order to enable the pull-ups make the relevant bit in PORTx high; however, if you do not want them make sure you disable them by making the relevant bit in PORTx low. For the outputs, we want to begin with the outputs in some sort of start state (e.g. all off), and so for the output pins, make the relevant bits in PORTx high or low depending on how you wish them to start. An example should clear things up.

Example 1.19   Using a 1200 chip, pins PB0, PB4 and PB7 are connected to push buttons. We would like pull-ups on PB4 and PB7 only. Pins PD0 to PD6 are connected to a seven segment display, and all other pins are not connected. All outputs should initially be off. What numbers should be written to DDRB, DDRD, PortB, and PortD to correctly specify the actions of the AVR's pins?

First, look at inputs and outputs. PB0, 4 and 7 are inputs, the rest are not connected (hence set as outputs). The number for **DDRB** is therefore **0b01101110**. For Port D, all pins are outputs or not connected, hence the number for **DDRD** is **0b1111111**.

To enable pull-ups for PB4 and PB7, make PortB, 4 and PortB, 7 high, all other outputs are initially low, so the number for **PortB** is **0b10010000**. All the outputs are low for Port D, so the number for **PortD** is **0b00000000**.

We can't move these numbers directly into the I/O registers, but instead we have first to move them into a working register (such as **temp**), and then output the working register to the I/O register. There are a number of ways we can do this:

> **ldi   register, number   ;**

This **load**s the immediate number into a register, but it is very important to note that this instruction cannot be used on all working registers – *only on those between R16 and R31* (we can therefore still use it on **temp**, as that is R16). We can also use a couple of alternatives to this instruction if the number we wish to move into the register happens to be 0 or 255/0xFF/0b11111111:

> **clr   register   ;**

This **cl**ears the contents of a register (moves 0 into it) – note an advantage of this over **ldi** is that it *can* operate on *all* working registers. Finally,

> **ser   register   ;**

This **se**ts the contents of a register (moves 255/0xFF/0b1111111 into it), though like **ldi**, it *only works on registers between R16 and R31*.

We then need to move **temp** into the I/O register, using the following instruction:

> **out   ioreg, reg**

This moves a number **out** from a register, into an I/O register. Make sure you note the order of the operands in the instruction – I/O register first, working register second, it is easy to get them the wrong way round! We can therefore see that the eight lines of the **Init** section move numbers into **DDRB**, **DDRD**, **PortB** and **PortD** via **temp**.

EXERCISE 1.13   Using a 1200 chip, pin PB0 is connected to a pressure sensor, and pins PB1, PB2 and PB3 control red, yellow and green LEDs respectively. PD0 to PD3 carry signals to an infrared transmitter, and PD4–PD6 carry signals from an infrared receiver. All other pins are not connected. All outputs should initially be off, and PB0 should have a pull-up enabled. Write the *eight* lines that will make up the **Init** section for this program.

After finishing the **Init** section, the program moves on to the main body of the program labelled **Start**. This is where the bulk of the program will lie. Note that the program ends with the line **rjmp   Start**. It needn't necessarily loop back to **Start**, but it does have to keep looping to *something*, so you may want to alter this last line accordingly. At the end of the program, you can write **.exit** to tell the assembler to stop assembling the file, but this isn't necessary as it will stop assembling anyway once it reaches the end of the file.

2
Basic operations with AT90S1200 and TINY12

The best way to learn is through example and by doing things yourself. For the rest of the book we will cover example projects, many of which will be largely written by you. For this to work most effectively, it helps if you actually try these programs, writing them out as you go along in Notepad, or whatever development environment you're using. If you don't have any special AVR software at the moment, you can still write the programs out in Notepad and test them later.

First of all, copy out the program template covered in the previous chapter, adjusting it as you see fit, and save it as **template.asm**. If you are using Notepad, make sure you select File Type as *Any File*. The **.asm** file extension refers to *assembly source*, i.e. that which will be assembled.

Program A: LEDon

● Controlling outputs

Our first few programs will use the 1200 chip. Load up the template, **Save As** to keep the original template unchanged, and call the file **ledon.asm**. Make the appropriate adjustments to the headers etc. relevant to the 1200 chip (header, **.device**, and **.include**). This first program is simply going to turn on an LED (and keep it on). The first step is to assign inputs and outputs. For this project we will need only one output, and will connect it to RB0. The second step in the design is the flowchart. This is shown in Figure 2.1. From this we can now write our program. The first box (Set-up) is performed in the **Init** routine. You should be able to complete this section yourself (remember, if a pin is not connected, make it an output).

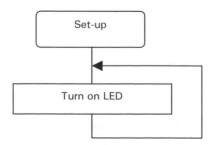

Figure 2.1

The second box involves turning on the LED, which means making RB0 high, which means setting bit 0 on PORTB to 1. To do this we could move a number into **temp**, and then move that number into **PortB**; however, there is a shortcut. We can use the following instruction:

> **sbi   ioreg, bit   ;**

This **s**ets a **b**it in an **I**/O register. Although you cannot move a number directly into an I/O register, you can set and clear the bits in *some* of them individually. You *cannot* set and clear individual bits in I/O registers 32–63 ($20–$3F in hex). Fortunately, PortB ($18) and indeed all the PORTx and PINx registers can be controlled in this fashion. The equivalent instruction for clearing the bit is:

> **cbi   ioreg, bit   ;**

This **c**lears a **b**it in an **I**/O register, though remember this only works for I/O registers 0–31. For our particular application, we will want to set **PortB, 0** and so will use the following instruction at the point labelled **Start**:

Start: sbi   PortB, 0   ; turns on the LED

The next line is:

> **rjmp   Start   ; loops back to Start**

This means the chip will be in an indefinite loop, turning on the LED. The program is now ready to be assembled. You can check that you've done everything right by looking at the complete program in Appendix J under *Program A*. All subsequent programs will be printed in the back in the same way. We will now assemble the program, but if you do not have the relevant software just read through the next section. You can download AVR Studio from Atmel's website (www.atmel.com) for free (last time I checked). This assembles, simulates and (with the right hardware) allows you to program the AVR chip.

AVR Studio – assembling

First of all load AVR Studio. Select **Project → New Project** and give it a name (e.g. LEDon), pick a suitable location, and choose AVR Assembler in the bottom box. In your project you can have assembly files, and other files. The program you have just written is an assembly file (.asm) and so you will have to add it to the project. Right click on **Assembly Files** in the Project Window and choose **Add File**. Find your original saved LEDon.asm and select it. You should now see your file in the Project Window. Now press **F7** or go to **Project → Assemble** and your file will be assembled. Hopefully your file should assemble with no errors. If errors are produced, you will find it helpful to examine the *List File* (*.lst). Load this up in Notepad, or some other text editor and scan the document for errors. In this simple program, it is probably nothing more than a spelling mistake. Correct any problems and then move on to testing.

Testing

There are three main ways to test your program:

1. Simulating
2. Emulating
3. Programming an actual AVR and putting it in a circuit

The first of these, *simulating*, is entirely software based. A piece of software pretends it's an AVR and shows you how it thinks the program would run, showing you how the registers are changing etc. You can also pretend to give it inputs by manually changing the numbers in PINB etc. You can get a good idea of whether or not the key concepts behind your program will work with this kind of testing, but other real-word factors such as button-bounce cannot be tested. Atmel's AVR Simulator comes with AVR Studio.

AVR Studio – simulating

We will now have a go at simulating the LEDon program. After you assemble your .asm file, double click on it in the Project Window to open it. Some of the buttons at the top of the screen should now become active. There are three key buttons involved in stepping through your program. The most useful one of these, 🔁 , is called Trace Into or Step Into. This runs the current line of your program. Pressing this once will begin the simulation and should highlight the first line of your program (**rjmp   Init**). You can use this button (or its shortcut **F11**) to step through your program. We will see the importance of the other stepping buttons when we look at subroutines later on in the book. In order for this simulation to tell us anything useful, we need to look at how the I/O registers are changing (in particular bit 0 of PortB). This can be done by going to **View → New IO View**. You can see that the I/O registers have been grouped into categories. Expand the PortB category and this shows you the PortB, DDRB and PinB registers. You can also view the working registers by going to **View → Registers**. We will be watching R16 in particular, as this is **temp**. Another useful shortcut is the reset button, 🔁 (**Shift + F5**).

Continue stepping through your program. Notice how **temp** gets cleared to 00, PortB and PortD are also cleared to 00, then **temp** is loaded with 0xFF (0b1111111), which is then loaded in DDRB and DDRD. Then (crucially) PortB, bit 0 is set, as shown by the tick in the appropriate box. You may notice how this will automatically set PinB, bit 0 as well. Remember the difference between PortB and PinB – PortB is a register representing what you wish to output through the port, and PinB represents the actual, physical state of those pins. For example, you could try to make an input high when the pin is accidentally shorted to ground – PortB would have that bit high whilst PinB would show the bit low, as the pin was being pulled low.

Emulating

Emulating can be far more helpful in pinning down bugs, and gives you a much more visual indication of the working of the program. This allows you to connect a probe with an end that looks like an AVR chip to your computer. The emulator software then makes the probe behave exactly like an AVR chip running your program. Putting this probe into your circuit should give you the same result as putting a real AVR in, the great difference being that you can step through the program slowly, and see the inner workings (registers etc.) changing. In this way you are testing the program and the circuit board, and the way they work together. Unfortunately, emulators can be expensive – a sample emulator is Atmel's ICE (In-Circuit Emulator).

If you don't have an emulator, or after you've finished emulating, you will have to program a real AVR chip and put it in your circuit or testing board. One of the great benefits of AVRs is the Flash memory which allows you to keep reprogramming the same chip, so you can quite happily program your AVR, see if it works, make some program adjustments, and then program it again with the new, improved code.

For these latter two testing methods you obviously need some sort of circuit or development board. If you are making your own circuit, you will need to ensure certain pins on the chip are wired up correctly. We will now examine how this is done.

Hardware

Figure 2.2 shows the 1200 chip. You will already be familiar with the PBx and PDx pins; however, there are other pins with specific functions. VCC is the positive supply pin, and in the case of the 1200 chip needs between 2.7 and 6.0 V. The allowed voltage range depends on the chip, but a value between 4 and 5 V is generally safe. GND is the ground (0 V) pin. There is also a $\overline{\text{Reset}}$ pin. The bar over the top means that it is *active low*, in other words to make the AVR reset you need to make this pin *low* (for at least 50 ns). Therefore, if we wanted a reset button, we could use an arrangement similar to that shown in Figure 2.3.

The power supply to the circuit is likely to take a short time to stabilize once first turned on, and crystal oscillators need a 'warm-up' time before they assume regular oscillations, and so it is necessary to make the AVR wait a short while after the power is turned on before running the program. Fortunately, this

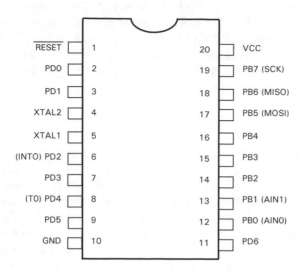

Figure 2.2

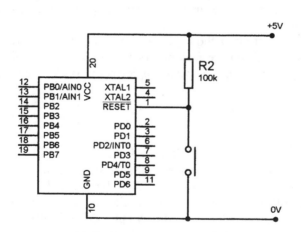

Figure 2.3

little delay is built into the AVR (lasting about 11 ms); however, if you have a particularly bad power supply or oscillator, and want to extend the length of this 'groggy morning feeling' delay you can do so with a circuit such as that shown in Figure 2.4. Increase the value of C1 to increase the delay.

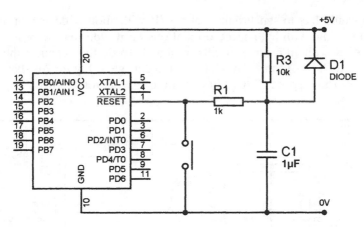

Figure 2.4

Finally, pins XTAL1 and XTAL2, as their names suggest, are wired to a crystal (or ceramic oscillator) which is going to provide the AVR with the steady pulse it needs in order to know when to move on to the next instruction. The faster the crystal, the faster the AVR will run through the program, though there are maximum frequencies for different models. This maximum is generally between 4 and 8 MHz, though the 1200 we are using in this chapter can run at speeds up to 12 MHz! Note that on some AVRs (in particular the Tiny AVRs and the 1200), there is a built-in oscillator of 1 MHz, which means you don't need a crystal. This internal oscillator is based on a resistor–capacitor arrangement, and is therefore less accurate and more susceptible to temperature variations etc.; however, if timing accuracy isn't an issue, it is handy to free up space on the circuit board and just use the internal oscillator. Figure 2.5 shows how you would wire up a crystal (or ceramic oscillator) to the two XTAL pins.

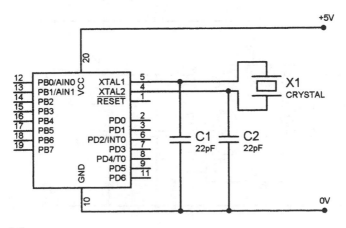

Figure 2.5

If you would like to synchronize your AVR with another device, or already have a clock line with high-speed oscillations on it, you may want to simply feed the AVR with an external oscillator signal. To do this, connect the oscillator signal to XTAL1, and leave XTAL2 unconnected. Figure 2.6 shows how using an HC (high-speed CMOS) buffer you can synchronize two AVR chips.

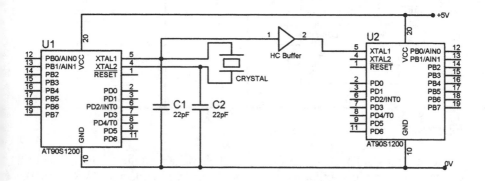

Figure 2.6

AVR Studio – programming

In order to test a programmed AVR, you will need a circuit board or development board. The simplest solution is to make up the circuit boards as you need them, but you may find it quicker to construct your own development board to cover a number of the projects covered in this book. The required circuit diagram for the LEDon program is shown in Figure 2.7.

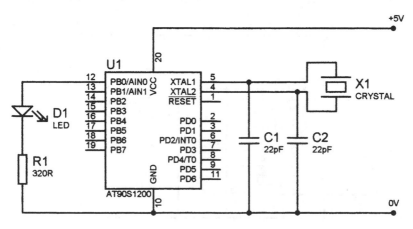

Figure 2.7

If you have a development board, you may need to check how the LEDs are wired up. We have been assuming the pins will *source* the LED's current (i.e. turn the pin high to turn on the LED). If your circuit board is configured such that the pin is *sinking* the LED's current, you will have to make changes to the software. In this case, a *0* will *turn on* the LED and a *1* will *turn off* the LED. Therefore, instead of starting with all of PortB set to 0 at the start of the Init section, you will want to move 0b11111111 into PortB (to turn off all the LEDs). You will also have to clear PortB, bit 0 rather than set it, in order to turn on the LED. This can be done using the **cbi** instruction in place of **sbi**.

Also note that although the program has been written with the 1200 in mind, by choosing the simplest model AVR we have made the program compatible with all other models (assuming they have sufficient I/O pins). Therefore if you have an 8515 (which comes with some development kits), simply change the **.device** and **.include** lines in your program and it should work.

We will now program the device using the STK500 Starter Kit. The steps required with the other types of programmer should not vary too much from these. To program your device, place the chip into the appropriate socket in the programming board. You many need to change the jumper cables to select the correct chip. In AVR Studio select **Tools** → **STK500**, and choose the relevant device (at90s1200). You will be programming the Flash Program memory. If you've just been simulating and your program is still in the simulator memory, you can tick the box labelled **Use Current Simulator/Emulator Flash Memory**, and then hit **Program**. If the program isn't in the Simulator/Emulator Memory, just load the program, assemble it, start the simulator, and it will be.

Fuse bits

You may notice some other tabs in the programming window. The one labelled fuses enables you to control some of the hardware characteristics of the AVR. These fuses vary between different models. For the 1200 we have two fuses available. **RCEN** should be set if you are using the internal RC oscillator as your clock. If you are using an external clock such as a crystal (as indeed we are in this project), this fuse bit should be clear. The other fuse is **SPIEN**, *Serial Program Downloading*, which allows you to read the program back off the chip. If you want to keep your program to yourself and don't want others to be able to read it off the chip, make sure this fuse bit is clear.

All this just to see an LED turn on may seem a bit of an anticlimax, but there are greater things to come!

Programs B and C: push button

- Testing inputs
- Controlling outputs

We will now examine how to test inputs and use this to control an output. Again, the project will be quite simple – a push button and an LED which turns on when the button is pressed, and turns off when it is released. There are two main ways in which we can test an input:

1. Test a particular bit in PINx using the **sbic** or **sbis** instructions
2. Read the entire number from PINx into a register using the **in** instruction

The push button will be connected between PD0 and 0V, and the LED to PB0. The flowchart is shown in Figure 1.3, and the circuit diagram in Figure 2.8.

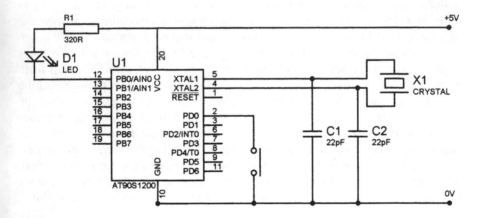

Figure 2.8

You should be able to write the **Init** section yourself, noting that as there is no external pull-up resistor shown in the circuit diagram, we need to enable the internal pull-up for PD0. The beginning of the program will look at testing to see if the push button has been pressed. We have two instructions at our disposal:

> **sbic   ioreg, bit   ;**

This tests a bit in a **I/O** register and skips the following line if the **bit** is **clear**. Similarly

> **sbis   ioreg, bit   ;**

tests a bit in a **I/O** register and skips the following line if the **bit** is **set**. Note that like **sbi** and **cbi**, these two instructions operate only on I/O registers numbered between 0 and 31 ($0–$1F). Fortunately, PIND, the register we will be testing, is one of these registers (number $10). So to test our push button (which makes pin PD0 high when it is pressed), we write:

sbis   PinD, 0   ; tests the push button

This instruction will make the AVR skip the next instruction if PD0 is high. Therefore the line below this one is only executed if the button is *not* pressed. This line should then turn off the LED, and so we will make the AVR *jump* to a section labelled **LEDoff**:

rjmp   LEDoff   ; jumps to the section labelled LEDoff

After this line is an instruction which is executed only when the button is pressed. This line should therefore turn the LED on, and we can use the same instruction as last time.

EXERCISE 2.1   Write the *two* instructions which turn the LED on, and then loop back to **Start** to test the button again.

This leaves us with the section labelled LEDoff.

EXERCISE 2.2   Write the *two* instructions which turn the LED off, and then loop back to **Start**.

You have now finished writing the program, and can double check you have everything correct by looking at Program B in Appendix J. You can then go through the steps given for testing and programming Program A. While you are doing your simulation, you can simulate the button being pressed by simply checking the box for **PIND, bit 0** in the I/O registers window.

Sometimes it helps to step back from the problem and look at it in a different light. Instead of looking at the button and LED as separate bits in the two ports, let's look at them with respect to how they affect the entire number in the ports. When the push button is pressed, the number in PinD is **0b00000000**, and in this case we want the LED to turn on (i.e. make the number in PortB **0b00000000**). When the push button isn't pressed, PinD is **0b00000001** and thus we want PortB to be **0b00000001**. So instead of testing using the individual bits we are going to use the entire number held in the file register. The entire program merely involves moving the number that is in PinD into PortB. This cannot be done directly, and so we will first have to read the number out of PinD using the following instruction:

in   register, ioreg   ;

This copies the number from an I/O register into a working register. To move the number from a working register back out to an I/O register, we use the **out** instruction. The entire program can therefore consist of:

Start:	**in**	**temp, PinD**	**; reads button**
	out	**PortB, temp**	**; controls LED**
	rjmp	**Start**	**; loops back**

This shorter program is shown as *Program C*.

Seven segment displays and indirect addressing

Using an AVR to control seven segment displays rather than using a separate decoder chip allows you to display whatever you want on them. Obviously all the numbers can be displayed, but also most letters: A, b, c, C, d, E, F, G, h, H, i, I, J, l, L, n, o, O, P, r, S, t, u, U, y and Z.

The pins of the seven segment display should all be connected to the same port, in any order (this may make PCB design easier). The spare bit may be used for the dot on the display. Make a note of which segments (a, b, c etc.) are connected to which bits. The segments on a seven segment display are labelled as shown in Figure 2.9.

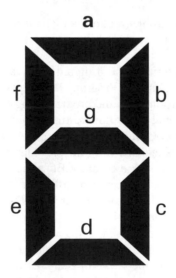

Figure 2.9

Example 2.1   Port B Bit 7 = d, Bit 6 = a, Bit 5 = c, Bit 4 = g, Bit 3 = b, Bit 2 = f, and Bit 1 = e. I have assigned the letters to bits in a random order to illustrate it doesn't matter how you wire them up. Sometimes you will find that due to physical PCB restrictions there are some configurations that are easier or more compact than others. The software is easy to change – the hardware normally less so.

If the display is wired up as described in Example 2.1, the number to be moved into Port B when something is to be displayed should be in the format **dacgbfe-** (it doesn't matter what bit 0 is as it isn't connected to the display), where the value associated with each letter corresponds to the required state of the pin going to that particular segment.

So if you are using a common cathode display (i.e. make the segments high for them to turn on – see Figure 2.10), and you want to display (for example) the letter **A**, you would turn on segments: a, b, c, e, f and g.

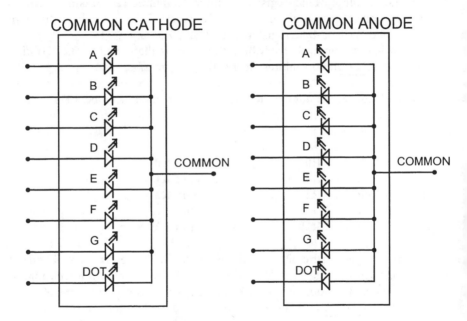

Figure 2.10

Given the situation in Example 2.1, where the segments are arranged **dacgbfe-** along Port B, the number to be moved into **PortB** to display an **A** would be **0b01111110**. Bit 0 has been made 0, as it is not connected to the display.

Example 2.2 If the segments of a common cathode display are arranged **dacgbfe-** along Port B, what number should be moved into **PortB**, to display the letter **C**, and the letter **E**?

The letter **C** requires segments a, d, e and f, so the number to be moved into Port B would be **0b11000110**. The letter **E** requires segments a, d, e, f and g so the number to be moved into Port B would be **0b11010110**.

EXERCISE 2.3 If the segments are arranged **abcdefg-** along Port B, what number should be moved into **PortB** to display the numbers 0, 1, 2, 3, 4, 5, 6, 7, 8, 9, A, b, c, d, E and F.

The process of converting a number into a seven segment code can be carried out in various ways, but by far the simplest involves using a *look-up table*. The key idea behind a look-up table is *indirect addressing*. So far we have been dealing with *direct addressing*, i.e. if we want to read a number from register number 4, we simply read register number 4. Indirect addressing involves reading a number from register number X, where X is given in a different register, called **Z** (the 2-byte register spread over R30 and R31).

It's a bit like sending a letter, where the letter is the contents of a working register (R0–R31), and the address is given by the number in **Z**.

Example 2.3 Move the number **00** into working registers numbers **R0** to **R29**.

Rather than writing:

clr	R0	; clears R0
clr	R1	; clears R1
clr	R2	; clears R2
etc.		
clr	R29	; clears R29

we can use indirect addressing to complete the job in fewer lines. The first address we want to write to is **R0** (address 0), so we should move **00** into **Z** (making 0 the address on the letter). **Z**, remember, is spread over both **ZL** and **ZH** (the higher and lower bytes of Z), so we need to clear them both:

clr	ZL	; clears ZL
clr	ZH	; clears ZH

We then need to set up a register with the number 0 so we can send it 'by post' to the other registers. We already have a register with a 0 (ZH), so we will use that.

st	register, Z	;

This indirectly **st**ores (sends) the value in **register** to the address pointed to by **Z**. Therefore the instruction:

st   ZH, Z   ;

sends the number in **ZH** (0) to the address given by **Z** (also 0), and so effectively clears **R0**. We now want to clear **R1**, and so we simply increment **Z** to point to address 01 (i.e. R1). The program then loops back to cycle through all the registers, clearing them all in far fewer lines that if we were using direct addressing. All we need to do is test to see when **ZL** reaches 30, as this is past the highest address we wish to clear.

How do we tell when ZL reaches 30? We subtract 30 from it and see whether or not the result is zero. If ZL *is* 30, then when we subtract 30 from it the result will be 0. We don't want to *actually* subtract 30 from ZL, or it will start going backwards fast! Instead we use one of the compare instructions:

cp   register, register   ;

This 'compares' the number in one register with that in another (actually subtracts one register from the other whilst leaving both unchanged). We then need to see if the result is zero. We can do this by looking at the *zero flag*. There are a number of flags held in the **SREG** register ($3F), these are automatically set and cleared depending on the result of certain operations. The zero flag is set when the result of an operation is zero. There are two ways to test the zero flag:

brbs   label, bit   ;

This **br**anches to another part of the program if a **b**it in SREG is **s**et (the zero flag is bit 1, and so **bit** would have to be a 1). Note that the label has to be within 63 instructions of the original instruction. Similarly,

brbc   label, bit   ;

This **br**anches to another part of the program if a **b**it in SREG is **c**lear. Here is where some of the instruction redundancy comes in, because as well as this general instruction for testing a bit in SREG, each bit has *its own particular instruction*. In this case, for the zero flag:

breq   label   ;

which stands for **br**anch if **eq**ual (more specifically, branch if the zero flag is set). The opposite of this is:

brne   label   ;

which stands for **br**anch if **n**ot equal (more specifically, branch if the zero flag is clear). The complete set of redundant/non-critical instructions is shown in Appendix C, along with their equivalent instructions. To compare a register with a number (rather than another register), we use the instruction:

cpi   register, number   ;

Please note that this only works on registers R16–R31, but as ZL is R30 we are all right. The complete set of instructions to clear registers R0 to R29 is therefore:

```
              clr    ZL            ; clears ZL
              clr    ZH            ; clears ZH
ClearLoop:    st     ZH, Z         ; clears indirect address
              inc    ZL            ; moves on to next address
              cpi    ZL, 30        ; compares ZL with 30
              brne   ClearLoop     ; branches to ClearLoop if ZL ≠ 30
```

This six line instruction set is useful to put in the **Init** subroutine to systematically clear a large number of file registers. You can adjust the starting and finishing addresses by changing the initial value of ZL and the final value you are testing for; note, however, that you don't want to clear ZL in the loop (i.e. don't go past 30) because otherwise you will be stuck in an endless loop (think about it).

EXERCISE 2.4   *Challenge!* What *six* lines will write a 0 to R0, a 1 to R1, a 2 to R2 etc. all the way to a 15 to R15?

As well as writing indirectly, we can also read indirectly:

ld   register, Z   ;

This indirectly **loads** into **register** the value at the address pointed to by **Z**. We therefore have a *table* of numbers kept in a set of consecutive memory addresses, and by varying **Z** we can read off different values. Say, for example, we keep the codes for the seven segment digits 0–9 in working registers R20–R29. We then move 20 into **Z** (to 'zero' it to point at the bottom of the *table*) and then add the number we wish to convert to **Z**. Reading indirectly into temp we then get the seven segment code for that number:

```
          ldi    ZL, 20         ; zeros ZL to R20
          add    ZL, digit      ; adds digit to ZL
          ld     temp, Z        ; reads Rx into temp
          out    PortB, temp    ; outputs temp to Port B
```

The above code translates the number in **digit** into a seven segment code which

is then outputted through Port B. Note that you will have to write the code to
the registers in the first place:

```
ldi     R20, 0b11111100     ; code for 0
ldi     R21, 0b01100000     ; code for 1
etc.
ldi     R29, 0b11110110     ; code for 9
```

Note that using working registers for this purpose is unusual and indeed
wasteful, but as there is no other SRAM on the 1200 we have no choice. On
other chips that do have SRAM, we can use that for look-up tables.
Furthermore, on other chips there is also an instruction **lpm**, which allows you
to use the Program Memory for look-up tables as well. More on this in the
Logic Gate Simulator project on page 67.

Programs D and E: counter

- Testing inputs
- Seven segment displays

Our next project will be a counter. It will count the number of times a push
button is pressed, from 0 to 9. After 10 counts (when it passes 9), the counter
should reset. The seven segment display will be connected to pins PB0 to PB6,
and the push button will go to PD0. Figure 2.11 shows the circuit diagram, pay
particular attention to how the outputs to the seven segment display are
arranged. The flowchart is shown in Figure 2.12.

You can write the Init section yourself, remembering the pull-up on the push
button. Start **PortB** with the code for a 0 on the display. We will be using a
register called **Counter** to keep track of the counts, you should define this in
the declarations section as **R17**. The reason we have assigned it R17 is that, as
you may remember, registers R16–R31 are the 'executive assistants' – more
powerful registers capable of a wider range of operations. We therefore tend to
fill up registers from R16 upwards, and then use R0–R15 if we run out. In the
Init section, set up registers **R20** to **R29** to hold the seven segment code for
numbers 0 to 9. (HINT: If you do this before setting up **PortB**, you can move
R20 straight into **PortB** to initialize it. Also remember to clear **Counter** in the
Init section.)

EXERCISE 2.5   What *three* lines will test the push button, loop back and test it
again if it isn't pressed? If it is pressed it should jump out of the loop and add
one to **Counter**?

Then we need to see whether **Counter** has exceeded 9. We use **cpi** to compare,
and **brne** to skip if they are not equal. If they are equal, **Counter** must be reset

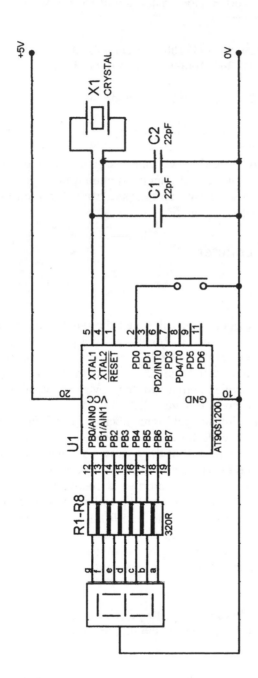

Figure 2.11

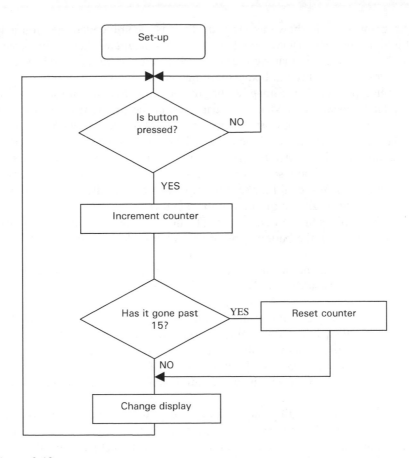

Figure 2.12

to 0. A useful trick with **brne** and similar instructions: it is often the case that rather than jumping somewhere exotic when the results aren't equal, we simply want to skip the next instruction (as we do with the **sbis** and **sbic** instructions). To do this with branch instructions, write **PC+2** instead of a label – this skips 1 instruction (i.e. jumps forward **2** instructions). PC stands for Program Counter which is described in more detail on page 54.

EXERCISE 2.6   What *three* lines will test if **Counter** is equal to 10 and reset it if it is? You may want to use the PC+2 trick.

Now we need to display the value in **Counter**. Do this by setting ZL to point to R20 and adding **Counter** to it, as described already.

EXERCISE 2.7   What *five* lines will display the value in **Counter** through Port B, and then loop back to **Start**?

The program so far is shown as Program D. It is recommended that you actually build this project. Try it out and you will spot the major flaw in the project.

The basic problem is that we are not waiting for the button to be released. This means that **Counter** is being incremented for the entire duration of the button being pressed. If we imagine that the button is held down for 0.1 s, and the crystal frequency is 4 MHz, one trip around the program takes about 14 clock cycles, and so **Counter** is incremented about $4\,000\,000/(14 \times 10) = 28\,600$ times for every press of the button! Effectively what we have is a pretty good random number generator (as an aside, random number generators are quite hard to make without some form of human input – computers are not good at being random). You could make this into an electronic dice project, but we will return to our original aim of a reliable counter.

Figure 2.13 shows the new flowchart. The necessary adjustment can be made at the end to wait for the button to be released before looping back to start.

EXERCISE 2.8   Write the *two* new lines needed to solve the problem, and show where they are to be added. (HINT: you will need to give this loop a name.)

Try out this new program (Program E), and you may notice a lingering problem, depending on the quality of your push button. You should see that the counter counts up in jumps when the push button in pressed (e.g. jumping up from 1 to 4). This is due to a problem called *button bounce*. The contacts of a push button actually bounce together when the push button is pressed or released, as shown in Figure 2.14.

In order to avoid counting one press as many, we will have to introduce a short delay after the button has been released before testing again. This affects the minimum time between counts, but a compromise must be reached.

Example 2.4   To avoid button bounce we could wait 5 seconds after the button has been released before we test it again. This would mean that if we pressed the button 3 seconds after having pressed it before, the signal wouldn't register. This would stop any bounce, but means the minimum time between signals is excessively large.

Example 2.5   Alternatively, to attempt to stop button bounce we could wait a hundred thousandth of a second after the button release before testing it again. The button bounce might well last longer than a hundred thousandth of a second so this delay would be ineffective.

A suitable compromise might be around a tenth of a second but this will vary from one type of button to the next and you will have to experiment a little. In order to implement this technique, we will have to learn about timing, which brings us to the next section.

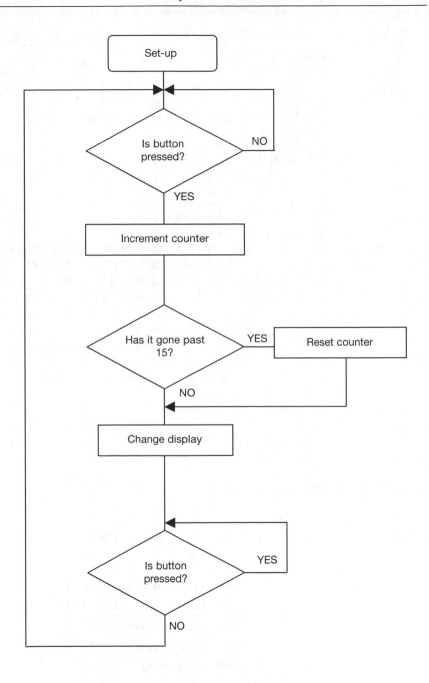

Figure 2.13

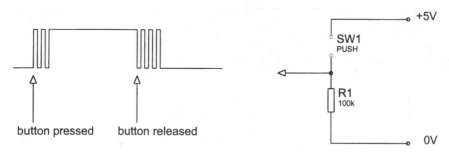

Figure 2.14

Timing

If you cast your mind back to the list of I/O registers (it may help if you glance back at page 14), you will notice a register called **TCNT0** ($32), or **T**imer **C**ounter **0**. This is an on-board timer, and will automatically count up at a specified rate, resetting to 0 when it passes 255. We can use this to perform timing functions (e.g. one second delays etc.). In more advanced chips there are several timers, some of which are 16 bits long. The reason it is also called a 'Counter' is that it can also be made to count the number of signals on a specific input pin (PD4 – pin 8 in the case of the 1200). For the purposes of the immediate discussion, we will be using **TCNT0** as a timer, and so I will be referring to it as **Timer 0**, or **T/C0** for the sake of brevity.

Before we can use Timer 0, we will have to configure it properly (e.g. tell it to time and not count). We do this with the **T/C0 C**onfiguration **R**egister: **TCCR0** ($33). In this register, each bit controls a certain aspect of the functioning of T/C0. In the case of the 1200, only bits 0–2 are used:

TCCR0 – T/C0 **C**ontrol **R**egister (**$33**)

bit no.	7	6	5	4	3	2	1	0
bit name	-	-	-	-	-	CS02	CS01	CS00

000	**STOP! T/C0 is stopped**
001	**T/C0 counts at the clock speed (CK)**
010	**T/C0 counts at CK/8**
011	**T/C0 counts at CK/64**
100	**T/C0 counts at CK/256**
101	**T/C0 counts at CK/1024**
110	**T/C0 counts on falling edge of T0 pin**
111	**T/C0 counts on rising edge of T0 pin**

Bits 3–7 have no purpose, but by setting bits 0–2 in a certain way, we can make T/C0 behave in the way we wish. If we don't wish to use T/C0 at all, all three bits should be 0. If we wish to use it as a timer, we select one of the next five options. Finally, if we want it to count external signals (on PD4), we can choose one of the last two options. The options available to us when using T/C0 for timing are to do with the speed at which it counts up. The clock speed (CK) is going to be very fast indeed (a few MHz) – this is the speed of the crystal which you connect to the AVR – and so in order to time lengths of the order of seconds we are going to have to slow things down considerably. The maximum factor by which we can slow down Timer 0 is 1024. Therefore if I connect a crystal with frequency 2.4576 MHz to the chip (this is actually a popular value crystal), Timer 0 will count up at a frequency of 2 457 600/1024 = 2400 Hz. So even if we slow it down by the maximum amount, Timer 0 is still counting up 2400 times a second.

Example 2.6   What number should be moved into the TCCR0 register in order to be able to use the T/C0 efficiently to eventually count the number of seconds which have passed?

Bits 3 to 7 are always 0.
Timer 0 is counting *internally*, at its slowest rate = CK/1024
Hence the number to be moved into the TCCR0 register is **0b00000101**.

EXERCISE 2.9   What number should be moved into the TCCR0 register when a button is connected between PD4 and +5 V, and TCNT0 is to count when the button is pressed.

In order to move a number into TCCR0, we have to load it into temp, and then use the **out** instruction, as with the other I/O registers. As you are unlikely to want to keep changing the Timer 0 settings it is a good idea to do this in the **Init** subroutine, to keep it out of the way.

   In order to time seconds and minutes, you need to perform some further frequency dividing yourself. We do this with what I call a *marker* and then any number of *counter* registers. These are working registers we use to help us with the timing. The basic idea is to count the number of times the value in Timer 0 reaches a certain number. For example, in order to wait one second, we need to wait for Timer 0 to count up 2400 times. This is equivalent to waiting for Timer 0 to reach 80, for a total of 30 times, because $30 \times 80 = 2400$. We could do this with any other factors of 2400 that are both less than 256.

   To test if the number in Timer 0 is 80, we use the following lines:

```
out     TCNT0, temp     ; copies TCNT0 to temp
cpi     temp, 80        ; compares temp with 80
breq    Equal           ; branches to Equal if temp = 80
```

This tests to see if Timer 0 is 80, and branches to Equal if it is. The problem is we're not always testing to see if Timer 0 is 80. The first time we are, but then next time round we're testing to see if Timer 0 is 160, and then 240 etc. We therefore have a register (which I call a marker) which we start off at 80, and then every time Timer 0 reaches the marker, we add another 80 to it. There isn't an instruction to add a number to a register, but there is one to subtract a number, and of course subtracting a negative number is the same as adding it.

        subi   register, number   ;

This **sub**tracts the immediate number from a register. *Note the register must be one of R16–R31.* So far, we have managed to work out when the Timer 0 advances by 80. We need this to happen 30 times for one second to pass. We take a register, move 30 into it to start with, and then subtract one from it every time Timer 0 reaches 80.

        dec   register   ;

This **dec**rements (subtracts one from) a register. When the register reaches 0 we know this has all happened 30 times. This all comes together below, showing the set of instructions required for a one second delay.

```
            ldi     Count30, 30     ; starts up the counter with 30
            ldi     Mark80, 80      ; starts up the marker with 80

TimeLoop:   out     TCNT0, temp     ; reads Timer 0 into temp
            cp      temp, Mark80    ; compares temp with Mark80
            brne    TimeLoop        ; if not equal keeps looping

            subi    Mark80, -80     ; adds 80 to Mark80

            dec     Count30         ; subtracts one from Count30
            brne    TimeLoop        ; if not zero keeps looping
```

The first two instructions load up the counter and marker registers with the correct values. Then TCNT0 is copied into temp, this is then compared with the marker. If they are not equal, the program keeps looping back to TimeLoop. If they are equal it then adds 80 to the marker, subtracts one from the counter, looping back to TimeLoop if it isn't zero. Note that you will have to define Mark80 and Count30 in the declarations section, and that they will have to be one of R16–R31.

Program F: chaser

● Timing
● Reading inputs
● Controlling outputs

The next example project will be a 'chaser' which consists of a row of LEDs. The LEDs are turned on in turn to give a chasing pattern. The speed of this chase will be controlled by two buttons – one to speed it up, the other to slow it down. The default speed will be 0.5 second per LED, going down to 0.1 second and up to 1 second.

The LEDs will be connected to Port B, and the buttons to PD0 and PD1. The flowchart and circuit diagram are shown in Figures 2.15 and 2.16 respectively.

The set-up box of the flowchart should be fairly straightforward, though remember that you may want to configure TCCR0 in the Init section, and that as we are timing the order of a second, we will want to use TCNT0 as a timer, slowed down by its maximum. Note also that PD0 and PD1 will require pull-ups, and that PortB should be initialized with one LED on (say, for example, PB0).

It is now worth giving a little thought to how we are going to have a time delay which can vary between 0.1 second and 1 second. The shortest time delay, 0.1 second, can be timed using a marker of 240 (2400/240 = 10 Hz), assuming the Timer 0 is counting at CK/1024 and a 2.4576 MHz crystal is being used. Then the counter can be varied between 1 and 10 to vary the overall time between 0.1 and 1 second. You may want to think about this a little. We will therefore have a marker register **Mark240**, and a variable counter register called **Counter**. **Counter** will be normally reset to 5 (for 0.5 second), but can be reset to other values given by **Speed**. Don't forget to define these registers at the declarations section at the top of the program).

Looking back at our flowchart, the first box after the set-up looks at the 'slow-down button'. We shall make the button at PD0 the 'slow-down button', and test this using the **sbic** instruction. If the button is not pressed (i.e. the pin is *high*), the next instruction will be executed, and this skips to a section where we test the 'speed-up button' button (call this **UpTest**).

If the button *is* pressed, we want to add one to **Speed** (slow down the chase). This can be done using the following instruction:

        inc   register   ;

This **inc**rements (adds one to) a register. We don't want the delay to grow longer than 1 second, and so we must check that **Speed** has not exceeded 10 (i.e. if it is 11 it has gone too far). We do this with the compare immediate instruction already introduced, **cpi**. If **Speed** is *not* equal to 11, we can then branch to **ReleaseDown** and wait for the button to be released. If it is equal to 11 we have

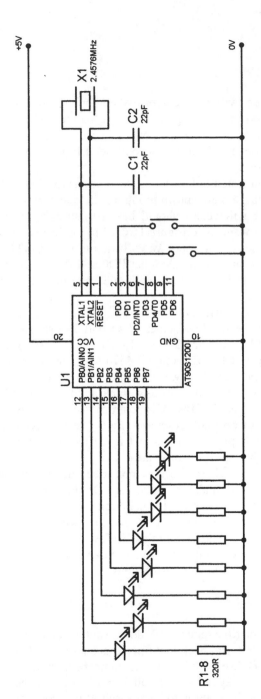

Figure 2.15

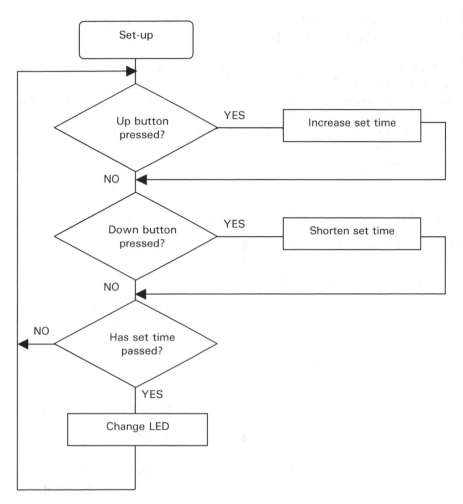

Figure 2.16

to subtract one from it (using the **dec** instruction). The first few lines of the program are therefore:

```
Start:        sbic    PinD, 0           ; checks slow-down button
              rjmp    UpTest            ; not pressed, jumps

              inc     Speed             ; slows down time
              cpi     Speed, 11         ; has Speed reached 11?
              brne    ReleaseDown       ; jumps to ReleaseDown if not equal
              dec     Speed             ; subtracts one from Speed
ReleaseDown:
```

```
        sbis    PinD, 0         ; waits for button to be released
        rjmp    ReleaseDown:    ;
```

In **UpTest**, we do the same with the 'speed-up button', PD1, and instead of jumping to **UpTest**, we jump to the next section which we will call **Timer**. If the speed-up button is pressed we need to decrement Speed, and instead of testing to see if it has reached 11, we test to see if it has reached 0 (and increment it if it has). We could use **cpi**   **Speed, 0**, but this line is unnecessary as the zero flag will be triggered by the result of the **dec** instruction, and so if we decrement Speed and the result is zero, we can use the **brne** in the same way as before.

EXERCISE 2.10   Write the *seven* lines which follow those given above.

The next section, called **Timer**, has to *check* to see if the set time has passed, and *return to the beginning if the time hasn't passed*. This means the timing routine must loop back to **Start** rather than stay in its own loop.

We will also put in the lines which set up the marker and counter registers in the **Init** section. **Mark240** should initially be loaded with 240; **Speed** and **Counter** should be loaded with 5. This means we can go straight into the counting loop.

```
Timer:  in      temp, TCNT0     ; reads Timer 0 into temp
        cp      temp, Mark240   ; compares temp with Mark240
        brne    Start           ; if not equal loops back to Start

        subi    Mark240, -240   ; adds 240 to Mark240

        dec     Counter         ; subtracts one from Counter
        brne    Start           ; if not zero loops back to Start
```

This should be familiar from the last section on timing. Note that instead of looping back to **Timer**, it loops back to **Start**. You may find, however, that you can reduce button bounce by looping back to **Timer** rather than **Start** in the 0.1 second loop. This means the buttons will only be tested once every 0.1 second, which means that a button will have to be pressed for at least 0.1 second. After the total time has passed, we need to chase the LEDs (i.e. rotate the pattern), and also reset the **Counter** register with the value in **Speed**. To do this we use:

```
        mov     reg1, reg2      ;
```

This **mov**es (copies) the number from reg2 into reg1.

EXERCISE 2.11   What *one* line resets **Counter** with the value in **Speed**?

To rotate the pattern of LEDs we have a number of rotating instructions at our disposal:

asr	**register**	; arithmetic shift right
lsr	**register**	; logical shift right
lsl	**register**	; logical shift left
ror	**register**	; rotate right
rol	**register**	; rotate left

The *arithmetic* shift right involves shifting all the bits to the right, whilst keeping bit 7 the same and pushing bit 0 into the *carry flag*. The carry flag is a flag in **SREG** like the zero flag. The *logical* shift right shifts all the bits to the right, and moves 0 into bit 7. The *rotate* right rotates through the carry flag (i.e. bit 7 is loaded with the carry flag, and bit 0 is loaded into the carry flag). This is summarized in Figure 2.17.

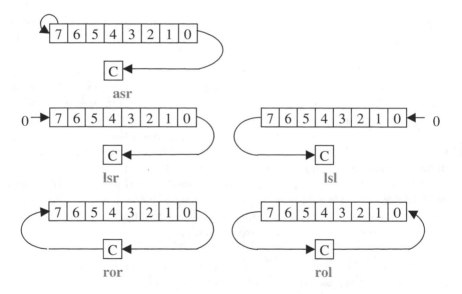

Figure 2.17

As we rotate the pattern along, we don't want any 1s appearing at the ends, because this would turn on edge LEDs out of turn, which would then propagate down the row and ruin the pattern. It would therefore seem that **lsl** or **lsr** is appropriate. For the sake of argument, we will pick **lsl**, to rotate the pattern to the left. We cannot apply these rotating instructions directly to **PortB**, so we have to read in the pattern to **temp**, rotate **temp**, and then output back to **PortB**. Before we output it to **PortB**, we have to see whether or not we've gone too far (rotated eight times), in which case we need to reset PortB back to its initial value (all off except PB0). We can do this by monitoring the carry flag, which will be high if we rotate a high bit off the end (a quick glance at Figure 2.17 should confirm this). The instruction for this is:

brcc   **label**   ;

This **br**anches to **label** if the **c**arry flag is **c**lear. Therefore the lines we need are:

```
in      temp, PortB         ; reads in current state
lsl     temp                ; rotates to the left
brcc    PC+2                ; checks Carry, skip if clear
ldi     temp, 0b00000001    ; resets to PB0 on, others off

out     PortB, temp         ; outputs to PortB
rjmp    Start               ; loops back to Start
```

You will notice that if the carry flag is clear, we skip the next instruction using the **PC+2** trick. The program is shown in its entirety as Program F in Appendix J.

You can go through and assemble this, and simulate it. For the simulation, you will notice that stepping through the entire program waiting for Timer 0 to count up will take a long time. For this reason, ways to run through parts of the program at high speed are on offer. For example, if you right click on a line in the program (when in simulation mode), you are given the option to 'Run to Cursor' (**Ctrl + F10**). This will run to where you have clicked at high speed (not quite real time, but close).

So far we have covered quite a few instructions; it is important to keep track of all of them, so you have them at your fingertips. Even if you can't remember the exact instruction name (you can look these up in Appendix C), you should be familiar with what instructions are available.

REVISION EXERCISE   What do the following do: **sbi, cbi, sbic, sbis, rjmp, ldi, st, ld, clr, ser, in, out, cp, cpi, brbs, brbc, breq, brne, brcc, subi, dec, inc, mov, asr, lsr, lsl, ror** and **rol**? (Answers in Appendix D.)

Timing without a timer?

Sometimes we will want to use the **TCNT0** for other purposes (such as counting signals on **T0/PD4**), and so we will now look at timing without the use of this timer. Each instruction takes a specific amount of time, so through the use of carefully constructed loops we can insert delays which are just as accurate as with Timer 0. The only drawback of this is that the loop cannot be interrupted (say, if a button is pressed), unlike the Timer 0, which will keep counting regardless.

The overall idea is to find the number of clock cycles we need to waste and count down from this value to 0. The problem lies when the number is greater than 255 (which is the case almost all the time). In this case we need to somehow split the number over a number of registers, and then cascade them. We decrement the lowest byte until it goes from 00 to FF (setting the carry flag as it does so), and then decrement the next highest byte etc.

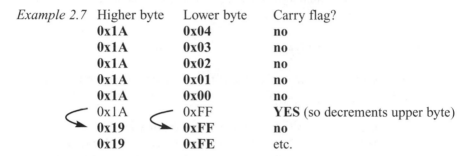

Example 2.7	Higher byte	Lower byte	Carry flag?
	0x1A	**0x04**	**no**
	0x1A	**0x03**	**no**
	0x1A	**0x02**	**no**
	0x1A	**0x01**	**no**
	0x1A	**0x00**	**no**
	0x1A	**0xFF**	**YES** (so decrements upper byte)
	0x19	**0xFF**	**no**
	0x19	**0xFE**	etc.

The first step is to work out how many instruction cycles the time delay requires. For example, to wait one second with a 4 MHz crystal, we need to 'kill' 4 million clock cycles. The loop we will write will take 'x' instruction cycles, where x is given in Table 2.1.

Table 2.1

x	**Length of time with 4 MHz clock**	**With 2.4576 MHz clock**
3	0–63 µs	0–102 µs
4	64 µs–16 ms	102 µs–26 ms
5	16 ms–4.1 seconds	26 ms–6.7 seconds
6	4.2 seconds–17 minutes	6.7 seconds–27 minutes
7	17 minutes–74 hours	27 minutes–120 hours

We are timing one second, which means $x = 5$. We therefore divide 4 000 000 by 5, getting in this case 800 000. We convert this number to hexadecimal, getting 0xC3500. Write this number with an even number of digits (i.e. add a leading 0 if there are an odd number of digits), and then split it up into groups of two digits. For example, our values are 0x00, 0x35 and 0x0C.

At the start of the delay in the program we put these numbers into file registers, note the order.

```
ldi      Delay1, 0x00        ;
ldi      Delay2, 0x35        ;
ldi      Delay3, 0x0C        ;
```

The delay itself consists of just one line per delay register plus one at the end (i.e. in our case four lines). To help us achieve such a short loop we need to use a new instruction:

> **sbci   reg, number   ;**

Subtract the immediate number from a register, and also subtract 1 if the carry flag is set. For example:

> **sbci   Delay2, 0   ;**

This effectively subtracts 1 from Delay 2 if the carry flag is set, and subtracts 0 otherwise. Our delay loop is as follows:

```
Loop:      subi    Delay1, 1   ; subtracts 1 from Delay1
           sbci    Delay2, 0   ; subtracts 1 from Delay2 if Carry is set
           sbci    Delay3, 0   ; subtracts 1 from Delay3 if Carry is set
           brcc    Loop        ; loops back if Carry is clear
```

When it finally skips out of the loop, one second will have passed. The first thing to note is that the length of the loop is five clock cycles (the branching instruction takes *two* clock cycles). You can now see where the numbers in Table 2.1 come from – for every extra delay register you add there is an extra cycle in the loop. The reason we have used **subi** to subtract 1 instead of **dec** is that unlike **subi**, **dec** doesn't affect the carry flag. We clearly rely on the carry flag in order to know when to subtract from the higher bytes, and when to skip out of the loop.

The program counter and subroutines

There is an inbuilt counter, called the *program counter*, which tells the AVR what instruction to execute next. For normal instructions, the program counter (or PC for short) is simply incremented to point to the next instruction in the program. For an **rjmp** or **brne** type instruction, the number in the PC is changed so that the AVR will skip to somewhere else in the program.

Example 2.8

```
Start:
039        sbi     PortB, 0    ; turns on LED
03A        sbic    PinD, 0     ; tests push button
03B        cbi     PortB, 0    ; turns off LED
```

Loop:

03C	**dec**	**Counter**	;
03D	**breq**	**PC+2**	; **skips next line if 0**
03E	**rjmp**	**Start**	;
03F	**rjmp**	**Loop**	;

The above example segment has the program memory addresses for each instruction on the left-hand side in hexadecimal. Note that blank lines aren't given addresses, nor are labels, for they are actually labelling the address that follows. Looking at the behaviour of the PC in the above, it starts at 039 and upon completion of the **sbi** instruction gets incremented to 03A. Then **PinD, 0** is tested. If it is high, the PC is simply incremented to 03B, but if it is low, the program skips, i.e. the PC is incremented twice to 03C. The **rjmp   Start** instruction moves 039 into the PC, making the program skip back to **Start**. This also sheds some light on the PC+2 trick we've used a few times already, if the result is 'not equal' (i.e. zero flag clear), the program adds 2 to the PC rather than 1, thus skipping one instruction.

EXERCISE 2.12   In the example above, what is the effect of the instruction **rjmp   Loop** on the PC?

This now brings us to the topic of *subroutines*. A subroutine is a set of instructions within the program which you can access from anywhere in the program. When the subroutine is finished, the program returns and carries on where it left off. The key feature here is the fact that the chip has to *remember* where it was when it called the subroutine so that it can know where to carry on from when it returns from the subroutine. This memory is kept in what is known as a *stack*. You can think of the stack as a stack of papers, so when the subroutine is called, the number in the program counter is placed on top of the stack. When a returning instruction is reached, the top number on the stack is placed back in the program counter, thus the AVR returns to execute the instruction after the one that called the subroutine. The 1200 has a *three level* stack. When a subroutine is called within a subroutine, the number in the PC is placed on top of the stack, *pushing* the previous number to the level below. The subsequent returning instruction will, as always, select the number on the top of the stack and put it into the PC. A three level stack means you can call a subroutine within a subroutine within a subroutine, but not a subroutine within a subroutine within a subroutine within a subroutine. This is because once you've pushed three values on to the stack, and you call another subroutine, hence pushing another value on to the stack, the bottom of the stack is lost permanently. The example in Figure 2.18 illustrates this problem.

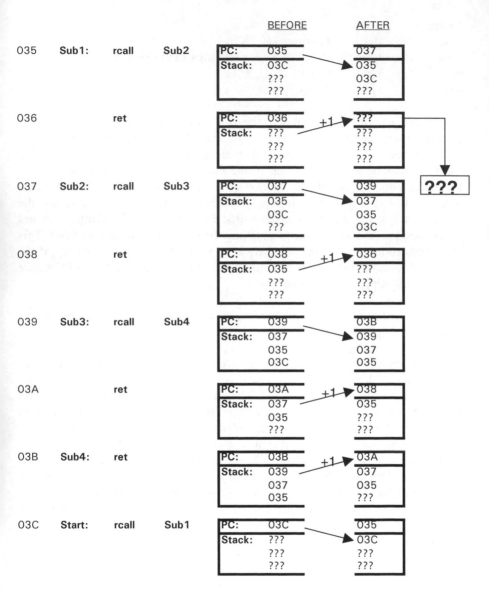

Figure 2.18

The instruction to call a subroutine is:

rcall   label   ;

Which is a **r**elative **call**, and so the subroutine needs to be within 2048 instructions of the **rcall** instruction. To **ret**urn from a subroutine use:

```
            ret                    ;
```

Of course, you can call as many subroutines as you like within the same subroutine like so:

```
Sub1:       rcall   Sub2           ;
            rcall   Sub3           ;
            rcall   Sub4           ;
            ret                    ;

Start:      rcall   Sub1           ;
```

Note that the programs so far have been upwardly compatible (this means they would work on more advanced types of AVR). This ceases to be strictly true with subroutines, and if you are developing these programs on a chip other than the 1200 or Tiny AVRs you will have to add the following four lines to the **Init** section – Chapter 3 explains why:

```
        ldi     temp, LOW(RAMEND)    ; stack pointer points to
        out     SPL, temp            ;    last RAM address
        ldi     temp, HIGH(RAMEND)   ;
        out     SPH, temp            ;
```

The simulator button 🠇 is used to *step over* a subroutine – i.e. it runs through the subroutine at high speed and then moves on to the next line. The *step out* button, 🠅, is used when the simulator pointer is in a subroutine and will make the simulator run until the return instruction is reached.

Program G: counter v. 3.0

● Debouncing inputs
● Seven segment display

Now that we know how to implement a timer, we can look back to improving the counter project to include debouncing features to counteract the effect of button bounce. The new flowchart is shown in Figure 2.19.

We can see from the flowchart that we need to insert two identical delays before and after the **ReleaseWait** section in the program. Rather than duplicating two delays, we can have a delay *subroutine* that we call twice. For example, if we call our delay subroutine **Debounce**, the following would be the last few lines of the new program:

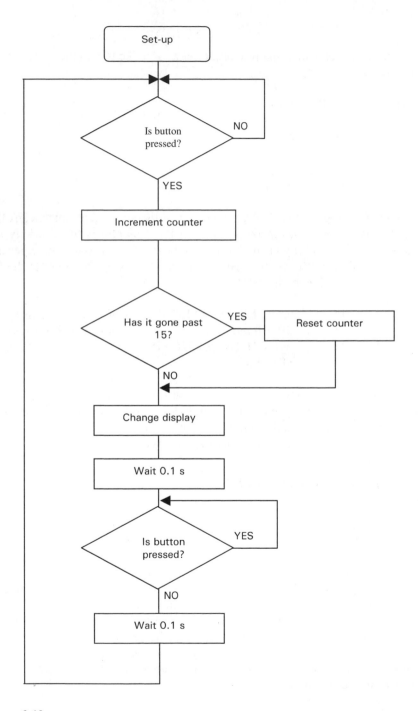

Figure 2.19

	rcall	Debounce	; inserts required delay
ReleaseWait:	sbis	PinD, 0	; button released?
	rjmp	ReleaseWait	; no, so keeps looping
	rcall	Debounce	; inserts required delay
	rjmp	Start	; yes, so loops back to start

Finally we can write the **Debounce** subroutine. I like to keep my subroutines in the top half of the page to keep things tidy, after the **rjmp   Init** line, but before the **Init** section itself. In this case we will use the delay without Timer 0.

EXERCISE 2.13   How many clock cycles will it take to create a 0.1 second delay, given a 4 MHz crystal? Convert this number into hexadecimal, and split it up over a number of bytes. What should the initial values of the delay registers be?

EXERCISE 2.14   *Challenge!* Write the *eight* lines that make up the **Debounce** subroutine.

You must also remember to define the three new registers you have added. With R20–R29 taken up by the seven segment code registers, and R30,31 belonging to ZL and ZH, you may think you've run out of useful room, and may have to use the less versatile R0–R15. However, notice that while in the **Debounce** subroutine, you are not using the **temp** register. You could therefore use **temp** instead of **Delay1**. Either define **Delay1** as R16 (there is nothing strictly wrong with giving a register two different names), or as this is potentially confusing you may prefer to scrap the name **Delay1** and use **temp** instead in the **Debounce** subroutine. Try this program out and see if you've eliminated the effect of the button bounce. Can you make the time delay smaller? What is the minimum time delay needed for reliable performance?

Program H: traffic lights

- Timing without Timer 0
- Toggling outputs

Our next project will be a traffic lights controller. There will be a set of traffic lights for motorists (green, amber and red), and a set of lights for pedestrians (red and green) with a yellow WAIT light as well. There will also be a button for pedestrians to press when they wish to cross the road. There will be two timing operations needed for the traffic lights. We will be monitoring the time between button presses as there will be a minimum time allowed between each time the traffic can be stopped (as is the case with real pedestrian crossings). As well as this, we will need to measure the length of time the lights stay on, and blinking. We will use the Timer 0 to control the minimum time between button presses (which we'll set to 25 seconds), and use the 'Timerless' method just introduced for all other timing. The circuit diagram is shown in Figure 2.20, and the flowchart in Figure 2.21.

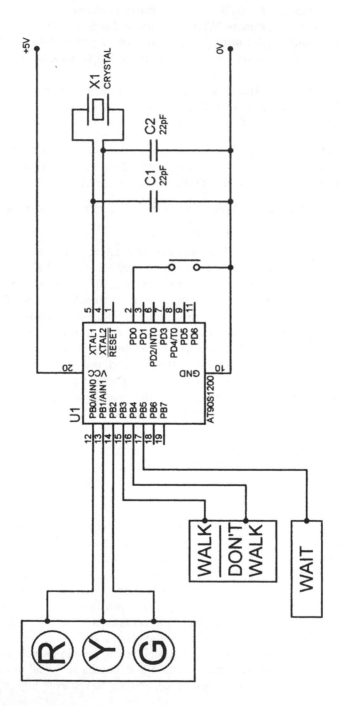

Figure 2.20

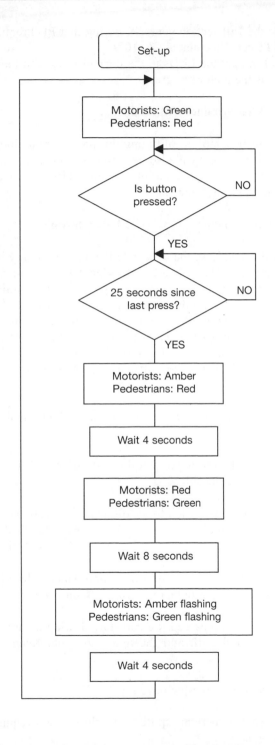

Figure 2.21

You can write the **Init** section yourself, noting that PD0 requires an internal pull-up. Set up **TCNT0** to count at CK/1024.

The first two lines get the LEDs in the correct state with the red pedestrian light on, as well as the motorists' green.

EXERCISE 2.15   What *two* lines will do this?

We need to perform some sort of timing during this initial loop so that while it is waiting for the button, it can also be timing out the necessary 25 seconds. This will be taken care of by a subroutine called **Timer** which we will write later. So after these two first lines insert:

> **rcall   Timer   ; keeps timing**

In this subroutine we will use the **T** bit in SREG, a temporary bit you can use for your own purposes. We will use it to signal to the rest of the program whether or not the required 25 seconds have passed. It will initially be off, but after the traffic is stopped, and the people cross etc., it is set. When it is set and **Timer** is called, it will count down, but rather that staying in a loop until the time has passed it returns (using **ret**) if the required time hasn't passed. When the required time does pass, the **T** bit is cleared again, and the rest of the program knows it's OK to stop the traffic again. After this instruction we test the button.

EXERCISE 2.16   What *two* lines will then test the push button and loop back to **Start** if it isn't pressed?

EXERCISE 2.17   If the button is pressed the pedestrian's WAIT light should be turned on, what *one* line does this?

To test the **T** bit, you can use one of the following instructions:

> **brts   label   ; br**anches if the **T** bit is **s**et
> **brtc   label   ; br**anches if the **T** bit is **c**lear

EXERCISE 2.18   What *two* lines form a new loop which calls **Timer**, and tests the **T** bit in SREG, staying in the loop until the **T** bit is clear.

After the required time has passed, we can start slowing the traffic down. Turn the green motorists' light off, and the amber one on. Keep all other lights unchanged.

EXERCISE 2.19   What *two* lines achieve this?

As the flowchart shows, there are quite a few time delays required, and rather than copy the same thing over and over, it makes sense to use a time delay subroutine. If we look at the minimum delay we will be timing (which is 0.5 second for the flashing), we can write a delay for this length and then just call it several times to create longer delays. The delay will be called **HalfSecond**, and so to wait 4 seconds we call this subroutine 8 times. We could simply write **rcall HalfSecond** eight times, but a shorter way would be the following:

```
            ldi      temp, 8       ;
FourSeconds:
            rcall    HalfSecond    ;
            dec      temp          ;
            brne     FourSeconds   ;
```

temp is loaded with 8, and then each time it is decremented, **HalfSecond** is called. After doing this eight times it skips out of the loop.

After this 4 second delay the red motorists' light must be turned on, and the amber one off. The red pedestrian light must be turned off, and the green one on. The pedestrian's WAIT light must also be turned off.

EXERCISE 2.20   Which *two* lines will make the required output changes?

EXERCISE 2.21   Which *four* lines make up an 8 second delay?

After the 8 seconds, the red motorists' light turns off, and the motorists' amber and pedestrians' green lights must flash. Start by turning the flashing lights on, and then we will look at how to make them flash.

EXERCISE 2.22   Which *two* lines make the required output changes?

To toggle the required two lights, we need to invert the states of the bits. There are two ways to invert bits. We could take the *one's complement* of a register, using:

```
            com      register       ;
```

This inverts the states of all of the bits in a register (0 becomes 1, 1 becomes 0).

EXERCISE 2.23   If the number in **temp** is **0b10110011**, what is its resulting value after **com temp**?

However, we want to *selectively* invert the bits. This is done using the *exclusive OR* logic command. A logic command looks at one or more bits (as its inputs) and depending on their states produces an output bit (the result of the logic operation). The table showing the effect of the more common *inclusive OR* command on 2 bits (known as a *truth table*) is shown below:

inputs		result
0	0	0
0	1	1
1	0	1
1	1	1

The output bit (**result**) is high if either the first **or** the second input bit is high (or if both are high). The exclusive OR is different in that if *both* inputs are high, the output is low:

inputs		result
0	0	0
0	1	1
1	0	1
1	1	0

One of the useful effects is that if the second bit is 1, the first bit is toggled, and if the second bit is 0, the first bit isn't toggled (see for yourself in the table). In this way certain bits can be selectively toggled. If we just wanted to toggle bit 0 of a file register, we would exclusive OR the file register with the number **00000001**.

The exclusive OR instruction is:

> **eor   reg1, reg2   ;**

This exclusive **OR**s the number in **reg2** with the number in **reg1**, leaving the result in **reg1**.

EXERCISE 2.24   What *four* lines will read state of the lights into **temp**, selectively toggle bits 1 and 3, and then output **temp** back to **PortB**. (Hint: You will need a new register, call it **tog**.)

EXERCISE 2.25   *Challenge!* Incorporate the previous answer into a loop that waits half a second, selectively toggles the correct lights, and repeats eight times. You will need a new register to count the number of times round the loop; call this **Counter**, and call the loop **FlashLoop**. This should take *eight* lines.

The traffic lights can now return to their original states, but before looping back to **Start**, remember to *set* the **T** bit. You can do this directly using the following instruction:

> **set**   ; sets the **T** bit

EXERCISE 2.26   Write the final *two* lines of the program.

What remains for us now are the two subroutines, **HalfSecond** and **Timer**. We will tackle **HalfSecond** first as it should be the more straightforward.

EXERCISE 2.27   Without using the Timer 0, create a half second delay, and use this to write the *eight* lines of the **HalfSecond** subroutine. A 2.4576 MHz crystal is being used.

For **Timer**, we first test the **T** bit. If it is clear we can simply return.

EXERCISE 2.28   Write the first *two* lines of the **Timer** subroutine.

We can then use the same method we used before in timing loops; however, instead of looping to the top of the section, return from the subroutine. The required time is 25 seconds, which on a 2.4576 MHz crystal with Timer 0 running at CK/1024 corresponds to a marker of 240 and a counter of 250 (work it out!).

EXERCISE 2.29   *Challenge!* Write the remaining *ten* lines of the **Timer** subroutine. Assume your counter and marker registers have been set up in the **Init** section (do this!), and reset the counter register with its initial value at the *end* of the subroutine. Don't forget to *clear* the **T** bit at the end of the subroutine (use the **clt** instruction).

Congratulations! You have essentially written this whole program yourself. To check the entire program, look at Program H (Appendix J).

Logic gates

We had a short look at the inclusive OR and exclusive OR logic gates, and now we'll look at other types: AND, NAND, NOR, ENOR, BUFFER, NOT. The truth tables are as follows:

AND

inputs		result
0	0	0
0	1	0
1	0	0
1	1	1

This is useful for *masking* (ignoring certain bits). If the second bit is 0, the first bit is *masked* (made 0). If the second bit is 1, the first bit remains intact.

Therefore ANDing a register with 0b00001111 masks bits 4–7 of the register, and leaves bits 0–3 the same.

NAND

inputs		result
0	0	1
0	1	1
1	0	1
1	1	0

This is the opposite of an AND

NOR

inputs		result
0	0	1
0	1	0
1	0	0
1	1	0

This is the opposite of an OR

ENOR

inputs		result
0	0	1
0	1	0
1	0	0
1	1	1

This is the opposite of an EOR

NOT

input	result
0	1
1	0

Only one input, output is opposite of input

Buffer

input	result
0	0
1	1

Only one input, output copies input

There aren't specific instructions for all these gates, but they can be implemented using a combination of available instructions.

Program I: logic gate simulator

- Logic functions
- TinyAVR

Our next project will be a logic gate simulator which can be programmed to act as any of the eight gates given above. It will therefore require two inputs and one output, and three inputs will together select which gate it is to emulate. This makes a total of six I/O pins, which just fits on the Tiny AVR chips. We will be writing this program for the Tiny12 AVR in particular, but it can be adapted to most of the other types, including the 1200 that we have so far been writing for. Figure 2.22 shows the pin layouts of some of the members of the Tiny family.

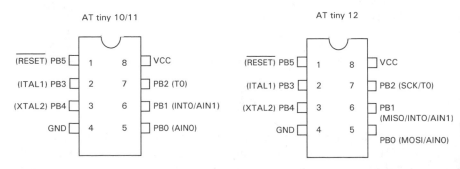

Figure 2.22

Basic features about this family include having a 6-bit Port B (PB0–PB5), but these six I/O pins are available only under certain circumstances. For example, you can see that PB3 and PB4 are also the oscillator inputs, and so to use these as I/O pins requires selection of the *internal* oscillator. Using a separate oscillator (and therefore only needing XTAL1 as a clock input) means PB4 *is* available, but PB3 isn't. Using the $\overline{\text{RESET}}$ pin as a reset pin means losing PB5. So you can see that having six I/O is very much a maximum. Also, take note that on the Tiny10 and Tiny11 PB5 is *an input only*. On the Tiny12, PB5 is an input or an *output drain* (this means you can make it an output, but only a low output – i.e. it can sink but not source current). This means that although **PinB** and **DDRB** are 6 bits long, **PortB** is only 5 bits long. PB5 therefore has no internal pull-up, and so needs an external resistor. An advantage of the Tiny AVRs over the 1200 model we have been using so far is the availability of the following instruction:

lpm   ;

This loads the contents of the program memory pointed to by **Z** into register R0. This means we can use the program memory itself as a look-up table, as opposed to using up working registers. It is also more efficient on code, as each instruction in the program memory is *16 bits* long, so we can store 2 bytes in place of an instruction. We will be needing this instruction in the example project.

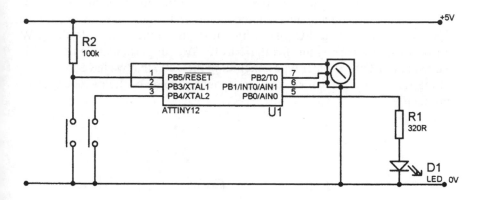

Figure 2.23

The circuit diagram for the logic gate project is shown in Figure 2.23. Note that the NOT and Buffer gates take only one input, and so we will be using PB1 as the input for these gates. Therefore, the effective *two-input* truth tables for the NOT and Buffer gates are:

NOT

inputs		result
0	0	1
0	1	1
1	0	0
1	1	0

Buffer

inputs		result
0	0	0
0	1	0
1	0	1
1	1	1

EXERCISE 2.30   Have a go yourself at constructing the flowchart, before looking at my version in the answer section. You need not make it more than three boxes in size, as we aren't yet concerned with sorting out how to manage the imitating of the individual gate types.

When writing the Init section the output, PB2, should initially be off. To choose which logic gate the AVR is to imitate, we have a binary switch which sets its outputs between (000) and (111) depending on the state of the switch. We therefore have to use this in the program to determine which section to jump to. Although the output from the switch is between 000 and 111, the resulting number in **PinB** is between xx000x and xx111x, where the states of bits 0, 4 and 5 must be ignored. We therefore take the number in **PinB** and mask bits 0, 4 and 5 using:

andi   reg, number   ;

This **AND**s the number in a register with the **immediate number** (*only for registers R16–R31*). To mask bits 0, 4 and 5, but keep bits 1–3 intact, we AND the register with 0b001110. We then rotate it once to the right, making sure that only zeros appear in bit 5 during the rotation.

EXERCISE 2.31   What is the appropriate rotation instruction to use?

The result is a number between 0 and 7 which we shall use to access a location in the program memory, and so we should load **PinB** into the **ZL** register as this will be used to point to a specific address.

EXERCISE *2.32*   Write the *three* lines which read **PinB** into **ZL**, mask bits 0, 4 and 5, and then rotate it to the right as required.

Our look-up table will begin after the **rjmp   Init** instruction. This instruction is at address 000 of the program memory (which is why it is the first one executed). Instructions are 16 bits long, and so take up 2 bytes (or one *word*). Program memory addresses are therefore *word addresses*, and the *byte address* is 2 times the word address. Figure 2.24 illustrates this.

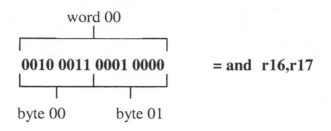

Figure 2.24

Our look-up table will therefore start at word address 001 which is equivalent to byte address 002. **ZL** points to the byte address, so we will have to add 2 to **ZL** to start it pointing to the bottom of the look-up table.

EXERCISE 2.33   Which *two* lines will add 2 to **ZL** and then use **ZL** to read a value from the program memory into R0?

Now the real question is what to have in the look-up table that is going to tell the program how to act like a particular logic gate. After some thought, I have found that using a split form of the truth table for each gate gives us the most straightforward solution. What we are about to do now may appear far from obvious, but hopefully after some thought you will see that ultimately it works rather neatly.

We are going to have a byte for each logic gate. For each gate, we take the truth table and look at the set of output states (e.g. 0001 for an AND gate, and 0111 for an inclusive OR). We then split these nibbles into two sets of 2 bits, and make these bits 4 and 5 and 0 and 1 of a byte. For example, AND: 0001 splits into 00 and 01, and then becomes 00000001. Inclusive OR: 0111 splits into 01 and 11, and the becomes 00010011.

EXERCISE 2.34   What are the relevant bytes for the NAND, NOR, ENOR, EOR, NOT and Buffer gates?

We then list these in the look-up table in any order we choose (noting that their position in the table defines how the code in PB1, 2 and 3 refers to a particular gate). The assembler has *directives* (instructions for the assembler) which tell it to place the following word or byte into the program memory. These directives are **.dw** (define **word**) and **.db** (define **byte**). If using **.dw**, you will have to group the bytes derived above into pairs (arbitrarily if you wish), e.g.:

.dw   0b0000000100010011   ; code for **AND** and **IOR**

OR

.db   0b00000001, 0b00010011 ; code for AND, code for IOR

There is one important difference between the two lines above. When using **.dw**, the lower byte of the word has the lower byte address. For example, if the two lines above were both written at *word address* 00, the code for the IOR would be at byte address 00 in the **.dw** example, and at byte address 01 in the **.db** example. As long as you take note of the correct byte addresses, it doesn't matter which way you do it.

EXERCISE 2.35   Complete the other *three* lines of the look-up table using **.dw** or **.db**.

Therefore, using the **lpm** instruction we have obtained a form of the truth table for each gate at **R0**. We will then test Input A of the gate (PB4). If it is low we *swap* the nibbles of **R0** (e.g. 00000001 becomes 00010000). What this does is select which half of the truth table we wish to access (remember we split it up into two halves). The swap instruction is:

>   **swap   reg   ;**

and **swap**s upper and lower nibbles of a register. We then test Input B of the gate (PB5). If it is low we *rotate* the number in **R0** to the right. What this does is select which of the two outputs remaining in the truth table is the right one. The four lines we need are therefore:

>   sbis   PinB, 4   ; tests Input A
>   swap   R0   ; swaps nibbles if low
>   sbis   PinB, 5   ; tests Input B
>   ror   R0   ; rotates right if low

The state of **R0, bit 0** now holds the output we wish to produce in PB0. However, we don't want to change the states of the pull-ups on the inputs, so we want to move a number into PortB that is all 1s for PB1–4, and PB0 equal to bit 0 of **R0**. Just like ANDing is a way to force certain bits low (masking), inclusive ORing is a way to force certain bits high. For example, in this case if we IOR **R0** with 0b11110 we will get a number that is all 1s except PB0 whose state is intact. We can then move the result of this into PortB safe in the knowledge that the pull-ups will remain. The inclusive OR instruction is:

>   **ori   reg, number   ;**

This inclusive **OR**s a register with the immediate **number**, but only works on registers R16–R31. We therefore have to move **R0** into **temp** using the **mov** instruction.

EXERCISE 2.36   What *four* lines take the number in R0, move it to **temp**, force bits 1–4 high and then output it to **PortB** before looping back to **Start**.

This finishes off the program, it is shown in its complete form in Appendix J.

SREG – the status register

We have seen some of the bits of SREG (zero flag, carry flag and T bit), and we will now look at the remaining five. They can all be individually tested, set or cleared using general SREG instructions: **brbc** and **brbs** which we have already met, and:

bset	**bit**	; **set**s a **bit** in SREG
bclr	**bit**	; **cl**ears a **bit** in SREG

Each bit also has its own personalized instructions (such as **breq** and **brcc**) which are listed in Appendix C. The bits in SREG are:

SREG – STATUS **Register ($3F)**

bit no.	7	6	5	4	3	2	1	0
bit name	I	T	H	S	V	N	Z	C

Carry flag:
Reacts to carrying
with arithmetic
operations, and to
the **ror** and **rol**
instructions.

Zero flag:
0: The result wasn't 0
1: The result was 0

Negative flag:
0: MSB of result is 0
1: MSB of result is 1

Two's complement overflow flag:
0: No two's complement overflow
1: Two's complement overflow
   occurred

Sign flag: (XOR of V and N bits)
0: Result is positive
1: Result is negative

Half carry flag:
Like the carry flag, except for the lower nibble
   (i.e. 4 lsbs)

T bit:
A temporary bit

Global interrupt enable:
Master switch for the interrupts
   (cleared when an interrupt occurs)

If you want to check whether a particular instruction affects a certain flag, check out the Instruction Overview (Appendix D). The purposes of the negative, two's complement overflow, and sign flags should be clear if you cast your mind back to the section on negative binary numbers. The half carry flag behaves in exactly the same way as the carry flag, except for the lower nibble. For example:

$$\begin{array}{r} 1111 \\ 01011010 = 90 \\ +\underline{00001111 = 15} \\ 01101001 = 105 \end{array}$$

This operation would set the half carry flag, as there was a carry on the bit 3 pair. The global interrupt enable will be introduced in the section on interrupts in Chapter 4.

Watchdog timer

A potentially useful feature of AVR chips is the *watchdog timer*: a 1 MHz internal timer, independent of outside components, which resets the AVR at regular intervals. In order to stop the AVR resetting, the watchdog timer must be cleared at regular intervals (i.e. before it has time to reset the chip). It is chiefly used as a safety feature, for if the program crashes the watchdog timer will shortly kick in and reset the chip, hopefully restoring normal operation. The watchdog timer is cleared using:

wdr   ;

This resets the watchdog timer (often called 'patting the dog'). The watchdog timer (WDT for short) is controlled by the WDTCR register:

WDTCR – Watchdog Timer Control Register ($21)

bit no.	7	6	5	4	3	2	1	0
bit name	-	-	-	-	WDE	WDP2	WDP1	WDP0

000	15 ms
001	30 ms
010	60 ms
011	0.12 second
100	0.24 second
101	0.49 second
110	0.97 second
111	1.9 seconds

Watchdog enable:
0: Watchdog Timer disabled
1: Watchdog Timer enabled

WDE controls whether or not the WDT is enabled, and WDP0-2 controls the length of time before the chip is reset. Note that the times given in the table are susceptible to temperature effects and are also a function of the supply voltage. The values in the table are for a supply of 5.0 V. For a 3.0 V supply the times are approximately three times longer.

Sleep

There are often applications where you wish the chip to be idle for a while until something happens. In such cases it is handy to be able to send the AVR to a low power mode called *sleep*. The AVR can be woken up from sleep by an external reset, a WDT reset, or by an interrupt (these are discussed in Chapter 4). The instruction to send the AVR to sleep is simply:

sleep ;

There are two types of sleep: a light snooze and a deep sleep. The light snooze (called *idle mode*) halts the program but keeps the timers (such as Timer 0) running. The deep sleep (called *power-down mode*) shuts down everything such that only the WDT, $\overline{\text{Reset}}$ pin, and INT0 interrupt can wake it up.

For example, to design a device that turns on when moved, we could do the following. Test the vibration switch and go to (deep) sleep if it is off. The WDT will then wake up the AVR and reset it. Testing the vibration switch will take a few microseconds, and the WDT could be set to time out every 60 ms, meaning the AVR is only on for about a thousandth of the time. When the vibration switch does eventually trigger the AVR will skip the sleep instruction and continue with normal operation. The WDT could then be disabled or reset at regular intervals using **wdr**.

To control the sleep properties of the AVR, we use an I/O register called **MCUCR ($35)**. Bit 5 of the **MCUCR** is the sleep enable, and this bit must be set if you wish to use the **sleep** instruction. Bit 4 selects which type of sleep you require (0 for idle mode and 1 for power-down mode).

More instructions – loose ends

Through the example projects we have encountered the majority of the instructions for the 1200 and Tiny AVRs. Here is the remainder:

> **neg   reg   ;**

This instruction makes the number in a register **neg**ative (i.e. takes the *two's complement*).

> **nop   ;**

This stands for **no op**eration, in other words do nothing. This essentially wastes one clock cycle, and can be quite useful. There are further instructions which perform logic and arithmetic operations on pairs of registers:

and	**reg1, reg2**	; **AND**s reg1 and reg2, leaving result in reg1
or	**reg1, reg2**	; **OR**s reg1 and reg2, leaving result in reg1
add	**reg1, reg2**	; **add**s reg1 and reg2, leaving result in reg1
adc	**reg1, reg2**	; as **add**, but adds an extra 1 if the **C**arry flag is set
sub	**reg1, reg2**	; **sub**tracts reg2 from reg1, leaving result in reg1
sbc	**reg1, reg2**	; as **sub**, but subtracts a further 1 if the **C**arry flag ; is set

There are also instructions to load a specific bit in a register into the **T** bit of SREG:

bst	**reg, bit**	; **st**ores a **b**it in a register into the **T** bit
bld	**reg, bit**	; **l**oads a **b**it in a register into the **T** bit

There are two more comparing instructions:

> **cpse   reg1, reg2   ;**

This **c**om**p**ares two registers and **s**kips the next instruction if they are **e**qual. In the same way that the **cp** instruction effectively performs a **sub** between two registers without actually changing them, the instruction **cpc** effectively performs an **sbc** between two registers without actually changing them. The SREG flags (e.g. carry and zero flag etc.) are affected in exactly the same was as with the **sub** and **sbc** instructions:

    cpc   **reg1, reg2**   ; compares two registers taking the Carry flag into
                          ;   account

Finally there are two instructions for testing the state of a bit in a working register:

    srbc   **reg, bit**   ; tests a bit in a register and skips next instruction if
                          ;   clear
    srbs   **reg, bit**   ; tests a bit in a register and skips next instruction if
                          ;   set

Major program J: frequency counter

- Multiple seven segment display
- Timing + counting
- Watchdog timer

We will end the chapter with a large project covering the key issues raised. We will design a frequency counter with a range 1 Hz–999 kHz. The frequency will be displayed on three seven segment displays, giving the frequency in Hz if it is less than 1 kHz, and in kHz otherwise. An LED will indicate the units. As an added feature, the device will stay on only when a signal greater than 1 Hz is fed into the input, and it will go to sleep when such a signal disappears. The circuit diagram is shown in Figure 2.25.

Notice that as we will be strobing the seven segment displays, each display will be on for only one-third of the time. In order to give each LED the same average current as it would be getting if it were being driven continuously, we need to divide the LEDs' series resistors by 3. Assuming a 5 V supply and a 2 V drop across the LED, there will be 3 V across the resistor. To supply a current of 10 mA to the LED if it were driven continuously, we would therefore choose a resistor value of 300 ohms. For this case I have therefore gone for a value of 100 ohms.

There are two ways to measure frequency. For high frequency signals it is best to take a fixed amount of time and count the number of oscillations on the input during that time. This can then be scaled up to represent a frequency. For lower frequency signals this method becomes too inaccurate, and so instead we measure the length of time between rising edges on the input. The program will have to work out whether the input frequency is high or low, and therefore which method it should use.

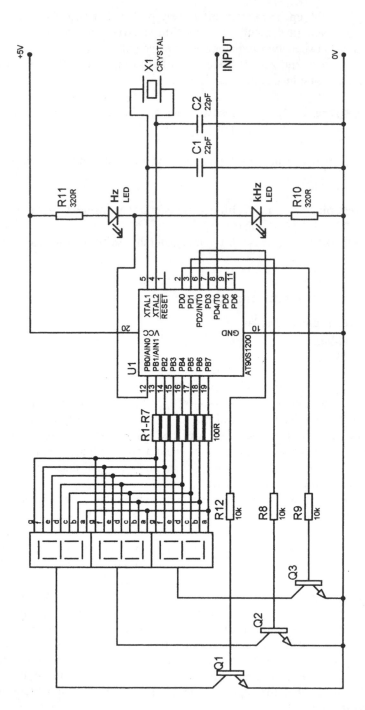

Figure 2.25

We have only one timer/counter at our disposal, which is an inconvenience, but something we can live with. For high frequency signals it is necessary to use T/C0 to count the input signal, as it will be difficult to test the input reliably. For lower frequency signals it will be easier to test the input directly, and more importantly to measure time accurately. This will be a long program, so it is all the more important to have a clear flowchart, shown in Figure 2.26.

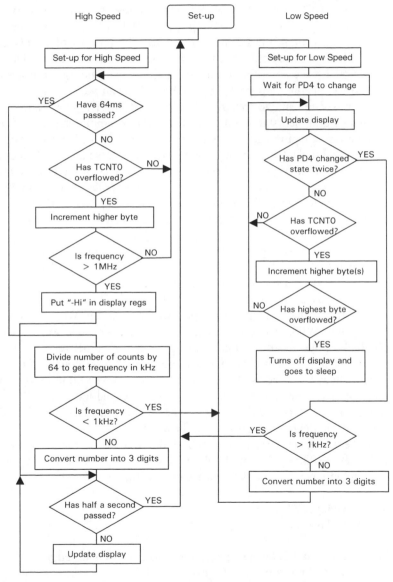

Figure 2.26

The test for high frequency signals takes the shortest time (64 ms), so the program will run this first. If the frequency measured is less than 1 kHz, the program will jump to the low-speed testing. The idea behind the high-speed testing is to time 64 ms by counting clock cycles (i.e. without T/C0), and count signals on T/C0. The only problem is that for timing up to 1 MHz, we would expect 64 000 cycles, i.e. well above 256. We therefore need to be monitoring T/C0 to see when it overflows, and increment a counter which would act as a higher byte for T/C0. You can now see why I chose 64 ms. The maximum number which can be stored over two registers is 0xFFFF = 65 536, so 64 000 is close to the maximum. Furthermore to convert the number of counts into a frequency in kHz, we need only to divide the number of counts by 64. Dividing a number by 2^n is easy – we simply rotate the number to the right n times (you may want to try this out on paper). This makes 64 ms an ideal choice.

For the low-speed test, we change T/C0 to count internally. We wait for the input to change and then start timing, waiting until the input changes a further two times before stopping again (this times the length of one cycle). Again, if we look at 1 Hz, with T/C0 counting at 4 MHz, this represents 4 million cycles, and we will need *three* registers to hold the entire number. If the time is greater than these three registers can hold, we know the time is less than 1 Hz, and so send the AVR to sleep. The WDT will be set to wake up the AVR every 1024 ms (i.e. about once a second), though note that in normal operation the WDT will have to be cleared regularly.

For the Init section, set up the ports with no pull-up on the input signal pin. Also, set up the WDTCR to enable the WDT to reset every 1024 ms, and configure MCUCR to enable *deep* (power-down) sleep.

We now need to carefully construct the main loop in which the timing will be carried out – this is the most important part of the program. We can guess that the loop is going to take somewhere between 4 and 10 cycles, so for 64 ms = 256 000 clock cycles, we are going to have to count down between 64 000 and 25 600 times, we can therefore make a guess that *two* counting registers (Delay1 and Delay2) can be used to count the time, but we will have to actually write the loop before we can be sure. Before we enter the loop we will have to set up the delay registers (we don't know what we will have to move into them as this depends on the loop length), set up how T/C0 is going to count, and reset T/C0 to 0. We will also use the move 0b10000000 into Port B to turn on the kHz LED and reset the display. You will notice there is also a line clearing a register called **upperbyte**, we will see the significance of this register shortly.

```
ldi     Delay1, ??          ;
ldi     Delay2, ??          ;

ldi     temp, 0b00000111    ; sets T/C0 to count rising edge
out     TCCR0, temp         ;    on T0 (PD4)
ldi     temp, 0b10000000    ; turns off all displays and turns on
```

```
out     PortB, temp          ;   kHz LED

clr     upperbyte            ; clears a counting register
clr     temp                 ; resets Timer 0
out     TCNT0, temp          ;
```

The loop itself starts with the standard decrementing of the 2-byte number spread over the delay registers, skipping out of the loop if the time has passed:

```
HighSpeed:
    subi    Delay1, 1        ; decrements Delay1
    sbci    Delay2, 0        ; decrements Delay2 if carry high
    brcs    DoneHi           ; jumps out of loop if time passed
```

We then need some way of testing to see if T/C0 has overflowed. There are two ways of doing this. The simplest is to test the *timer overflow flag*, which, unlike the other flags we've met so far, is stored in the **TIFR** I/O register. Unfortunately, we cannot test this flag directly with the **sbic** or **sbis** instructions, as it is number 0x38 which is greater than 0x1F. We would therefore have to read TIFR into a working register, then test the bit. More irritating is the fact that we need to reset it by writing a one to it. Again, we cannot use the **sbi** instruction, and instead have to do it through a working register. This overall process takes five instructions, but there is an alternative method which only uses four. The concept behind this method is to store the current value of T/C0 and compare it with the value that was in T/C0 the previous time in the loop. We would expect the current value to always be greater than the previous value, *except when it overflows*. By comparing the old and new values, and branching if the new is less than the old, we therefore detect an overflow, and no resetting of flags is needed. In the code below, we use register **temp** to store the new value, and **temp2** to store the old value:

```
mov     temp2, temp          ; copies temp into temp2 (old value)
in      temp, TCNT0          ; reads new value into temp
cp      temp, temp2          ; compares old and new
brsh    HighSpeed            ; loops back if new is 'same or higher'
```

If you count through the total **HighSpeed** loop of seven instructions, you will see it takes eight clock cycles if T/C0 doesn't overflow (remember a branching instruction takes two clock cycles). What we need to do now is construct a similar loop that will increment the higher byte, see if it's too high, decrement our counting registers, skip out if they've reached zero, and loop back to **HighSpeed**, all *in the same number of clock cycles*. This final part is crucial to ensure the timing is perfect. Fortunately we can do it all, with a clock cycle to spare! We therefore use **nop** to waste one cycle. The maximum number of counts we are allowing on the input is 63 999 in the 64 ms (i.e. 1 MHz is just too high, and so 64 000 is just too high – 64 000 translates as 0xFA00, which is handy as we can simply test if the upper byte has reached 0xFA). If it has we know how to skip out of the loop:

```
inc     upperbyte          ; increments higher byte
cpi     upperbyte, 0xFA    ; too high?
breq    TooHigh            ; skips out of loop if too high
subi    Delay1, 1          ; decrements counting registers
sbci    Delay2, 0          ;
brcs    DoneHi             ; skips out of loop if done counting
nop                        ; wastes one cycle
rjmp    HighSpeed          ; loops back
```

Now you may be thinking 'hang on, there are *nine* cycles in the above segment, not eight!'. You are right, of course, but think about the number of cycles in the previous section if the program does *not* loop back to **HighSpeed**. If the program does *not* loop back, it does *not* branch, and so takes one less clock cycle. We make up for this one less clock cycle in the loop above with one more in this loop. Thus in the running of this whole section, the counting registers will either decrement once every eight clock cycles or twice every 16 clock cycles. You may want to write the whole loop down and work through it to convince yourself of this. Now that we know the delay registers decrement every eight clock cycles, we can work out what to initialize them to in order to create a 64 ms delay.

EXERCISE 2.37   What should Delay1 and Delay2 be initialized to?

That's the hardest part done! We now need to immediately store the current value of T/C0. The only problem is, what if T/C0 has overflowed in between the last test for overflowing and now? We need to use the same test as before.

EXERCISE 2.38   Write the *six* lines which make up the section called **DoneHi**, which stores T/C0 into **lowerbyte**, and compare this value with **temp** (which represents the old value of T/C0). If **lowerbyte** is 'same or higher' it skips to a section called **Divide64**, if it isn't, it increments **upperbyte**, tests to see if it has reached 0xFA, and jumps to **TooHigh** if it has.

The next section needs to divide the 2-byte number split up over **lowerbyte** and **upperbyte** by $64 = 2^6$. We do this by rotating the whole number six times; to rotate the upper byte into the lower byte, we rotate the upper byte right with zeros filling bit 7, and then rotate the lower byte right with the carry flag filling bit 7.

EXERCISE 2.39   What *two* lines divide the 2-byte number by 2?

The **Divide64** loop does this six times. First we set up **temp** with the number 6, then divide by 2 as we've done above. Then decrement **temp**, looping back if it does not equal zero. We don't want to reset **temp** with 6, so we really want to jump to **Divide64** and then skip one instruction. This can be done using the following trick:

rjmp   Divide64+1   ; jumps to Divide64 and then skips one

This works with any jumping/branching instruction, and for any number of skips. Note that large skips (e.g. **+8**) lead to unwieldy programs which are hard to follow and easy to get wrong.

EXERCISE 2.40   What *five* lines make up the **Divide64** section?

We test to see if the number is too low. The 2-byte word holds the frequency in kHz, so if this number is less than 1 (i.e. 0) we know how to change to the low-speed testing method.

EXERCISE 2.41   What *four* lines test to see if both bytes are 0, and skips to **LowSpeed** if they are.

We then need to convert this number split over 2 bytes into a number of hundreds, tens and ones so that they can be displayed easily. This will be done in a subroutine, as we will have to do it in the **LowSpeed** section as well. To do the conversion we will call **DigitConvert**. As the displays are being strobed, we need to be calling a display subroutine at regular intervals. Unfortunately, our carefully constructed timing loop above cannot accommodate the calling of a display subroutine, as this would insert large numbers of clock cycles and disrupt the timing. The timing routine only takes 64 ms, so the idea here is to leave the displays idle for 64 ms, and then let them run for half a second.

We stick in a simple half second delay using counting registers, making sure we call the **Display** subroutine during the loop.

EXERCISE 2.42   Write the *nine* instructions which set up the three delay registers, and then create a half second delay loop which also calls **Display**. When the required time has passed, the program should jump back to **Start**. You will have to take the length of the **Display** subroutine into account when doing your calculations. The **rcall** instruction actually takes three cycles, and the **ret** instruction takes four. On average, the subroutine itself will take two instructions, so assume the whole subroutine action adds nine clock cycles to the loop. Call the delay loop **HalfSecond**.

All that remains in the high-speed timing method is to deal with the **TooHigh** section, which simply has to make the display registers show –HI. The numbers to be displayed will be stored in registers called **Hundreds**, **Tens** and **Ones**. There will be a look-up table as before, except in this table 10 will be translated as the symbol for an 'H', and 11 as the symbol for a hyphen '-'. A 12 will be translated as a blank space (i.e. no segments on), and so you should set all digits to 12 in the Init section. We therefore need to move 11 into **Hundreds**, 10 into **Tens** and a 1 into **Ones** (as a 1 will look like an I), and the **Display** subroutine will do the rest. After this we jump to *three lines before* the start **HalfSecond** section (these three lines previously set up the **HalfSecond** counting registers).

EXERCISE 2.43   What *four* lines make up the **TooHigh** section?

This marks the end of the high-speed timing method, and therefore the halfway point in the program.

Let's have a look at the **DigitConvert** subroutine. This takes a number split over **upperbyte** and **lowerbyte**, and converts it into a number of hundreds, tens and ones. This is done by repeatedly subtracting 100 from the 2-byte number until there is a carry. 100 is then added back, and the process is repeated with 10. The number left in the lower byte after this is simply the number of ones, so we can just move the number across. Once we have extracted the number of hundreds, we no longer need to involve the upper byte, as we know the number is now entirely contained in the lower byte (if the number is less than 100 it fits in one byte).

DigitConvert:

```
              clr     Hundreds        ; resets registers
              clr     Ones            ;
              clr     Tens            ;
```

FindHundreds:

```
              subi    lowerbyte, 100  ; subtracts 100 from lower byte
              sbci    upperbyte, 0    ; subtracts 1 if carry
              brcs    FindTens        ; does 10's if carry
              inc     Hundreds        ; increment number of hundreds
              rjmp    FindHundreds    ; repeats
```

FindTens:

```
              subi    lowerbyte, -100 ; adds back the last 100
              subi    lowerbyte, 10   ; subtracts 10 from lower byte
              brcs    FindOnes        ; does 1's if carry
              inc     Tens            ; increments number of tens
              rjmp    FindTens+1      ; repeats, but doesn't add 100 again
```

FindOnes:

```
              subi    lowerbyte, -10  ; adds back the last 10
```

```
mov      ones, lowerbyte   ; number left in lowerbyte = ones
ret                        ; finished
```

You may want to work your way through this program with a sample number (e.g. convince yourself that 329 gets reduced to 3 hundreds, 2 tens and 9 ones). The other subroutine is **Display**. This has to choose which of the three displays to activate, find the appropriate number in **Hundreds**, **Tens** or **Ones**, and then display it. In the half second loop we've written, the subroutine is called about once every 4 ms. We can't make the displays change this often as the LEDs won't have time to turn fully on and the display will be faint with shadows (numbers on other displays appearing on the wrong display). We therefore build in an automatic scaling of 50 – i.e. the subroutine returns immediately having done nothing 49 times, and then on the 50th time it's called, it performs the display routine, and then repeats. This means the displays are changing every 0.2 ms which is far better; however, should you experience any of the effects described above, you may wish to increase 50 to a higher value.

We will use a register called **DisplayCounter**. This will be set up in the Init section with the value 50. The beginning of **Display** therefore decrements **DisplayCounter**, and returns if the result is not 0. If it is 0, **DisplayCounter** should be reloaded with 50. Furthermore, we can take this opportunity to clear watchdog timer. This must be done regularly, and the **Display** subroutine is called regularly in whichever part of the program it happens to be (by regularly I mean at least once a second). A simple solution is therefore to reset the WDT when the **Display** subroutine continues.

EXERCISE 2.44 Write the *five* lines at the start of the **Display** subroutine.

We need some way to know which display we will be displaying, and will store this as a number between 0 and 2 in a register called **DisplayNumber**. Therefore, the first thing we do is increment **DisplayNumber** and reset it to 0 if it has reached 3 (you will also have to clear **DisplayNumber** in the Init section).

EXERCISE 2.45 Write the subsequent *four* lines of the subroutine which perform this.

Now we need to do some serious indirect addressing! First, we extract the right number to be displayed from **Hundreds**, **Tens** or **Ones**. You will have to define these at the top of the program, I defined mine as R26, R27 and R28 respectively. We therefore set up ZL to point to R26 (move 26 into ZL), and then add the number in **DisplayNumber**. This will point to one of the three numbers we want to display. Using the **ld** instruction we load this value into **temp**. The seven segment display codes are stored in registers R0–R12, and so we now zero ZL to R0 (move 0 into it). Adding to R0 the number read into **temp** should point to the seven segment code of the number to be displayed. Again, load this value into **temp**. We mustn't clear bit 7 of PortB if it is on (indicating kHz). Therefore, test bit 7 of Port B, if it is on, OR the number in **temp** with 0b10000000, and then in either case move **temp** into Port B.

EXERCISE 2.46   Write the *nine* lines which output the correct seven segment code to Port B.

The remainder of the subroutine must turn on the correct seven segment display. Remember the essence of strobing: the number you have just outputted to Port B is going to *all* of the displays, but by turning only one of them on, the number only appears in one of the displays. We basically want to turn on PortD bit 0, then bit 1, then bit 2 and then back to bit 0. The easiest way to do this is to read PinD into **temp**, rotate it left without letting any 1s creep in (i.e. use **lsl**), test to see if bit 3 is high (i.e. gone too far), and reset the value to 0b00000001 if it is.

EXERCISE 2.47   What *six* lines turn on the correct display and then return from the subroutine?

Now all that is left is the low-speed testing section. We need to set up T/C0 to count up every clock cycle (this gives us maximum resolution). We also need to (reset) clear the delay registers **Delay2** and **Delay3**, and clear PB7 to turn on the Hz LED.

EXERCISE 2.48   What *five* lines will start off the **LowSpeed** section?

We need a way to see when PD4 changes (remember now T/C0 is counting internally we need to test the input pin manually). There are a few methods at our disposal, the one I suggest is as follows. Store the initial value in PinD, and then enter a loop which reads in the current value of PinD, and *exclusive OR* it with the initial value. The effect of the EOR is to highlight which bits are different.

Example 2.9   0b00011001
                    EOR 0b10001001
                       0b10010000   ← shows that bits 7 and 4 were different

We are interested only in bit 4 (PD4) which is connected to the input, and so after performing the EOR we can test bit 4 of the answer and keep looping until it is high. When in any loop that lasts a long time (as this one might), we must also keep calling the **Display** routine.

```
                in       store, PinD    ; stores initial value
FirstChange: rcall       Display        ; keeps displays going
```

```
in      store2, PinD    ; reads in current value
eor     store2, store   ; EORs current and initial values
sbrs    store2, 4       ; skips out of loop if PD4 changed
rjmp    FirstChange     ; keeps looping until PD4 changes
```

The main loop of the low-speed testing section consists of repeating the above test for *two* changes (i.e. wait for one complete period of the input's oscillation), and incrementing the higher bytes when T/C0 overflows. We deal with the T/C0 overflow in the same way as before, with one important difference. We cannot use **temp** to store the old value because **temp** is used repeatedly in the **Display** subroutine we have just written. It is very important you look out for these kinds of traps as they can be a source of many problems – try to keep your use of working registers local (i.e. don't expect them to hold a number for too long), in this way you can use a register like **temp** all over the program. We can use **Delay1** instead of **temp**, as at the end of the looping, we want **Delay1** to hold the current value in T/C0.

Before we enter the low-speed loop we need to clear **Delay1** and T/C0. We will also need some sort of counter to count the number of times the input changes. We need it to change only twice, so set up a register called **Counter** and load 2 into it.

EXERCISE 2.49   Write the *three* pre-loop instructions.

Now the loop looks for a change in the input in the same way as before, and jumps to a section called **Change** if there is a change.

EXERCISE 2.50   Write the *five* lines which perform this test. (HINT: One of them is before the start of the loop, call the loop **LowLoop**.)

We then call the **Display** subroutine, as we have to do this regularly, then test to see if the T/C0 has overflowed. If it hasn't overflowed, loop back to **LowLoop**. If it has overflowed, increment **Delay2**, and if this overflows increment **Delay3**. The minimum frequency is 1 Hz, and hence the maximum amount of time is about 4 000 000 counts, which in hexadecimal is 0x3D0900. Therefore if **Delay3** reaches 0x3E we know the input frequency is too slow and will jump to a section called **TooSlow**.

EXERCISE 2.51   *Challenge!* What *11* lines form the rest of the low-speed section.

The **Change** section should decrement **Counter**, and loop back to **LowLoop** if it isn't zero. On the second change, it doesn't loop back but instead checks to see if the stored number is low enough to deserve high-speed testing. The maximum frequency measured with this method is 999 Hz, which corresponds to 4004 clock cycles, hence if the result is 4000 (0xFA0) or less we should branch to **Start** and perform the high-speed testing. It may not be entirely clear how we test to see if the number spread over three registers is less than 0x000FA0. For a start, we cannot subtract the number, as this would change the number in the delay registers. Instead, we use the compare instructions as we would if we were just testing one byte, but also make use of the **cpc** instruction, which compares two registers and also takes the carry flag into account. It is simply analogous to subtracting with the carry (e.g. **sbci** but without actually changing the registers). The only problem with **cpc** is that it only works between two file registers, not a file register and a number, so we have to load the numbers into temporary working registers. The necessary lines for **Change** are therefore:

```
Change:  in     store, PortB    ; updates new value of PortB
         dec    Counter         ; waits for second change
         brne   LowLoop         ; not second change so loops

         ldi    temp, 0x0F      ; sets ups temporary registers
         ldi    temp2, 0x00     ;
         cpi    Delay1, 0xA0    ; compares three-byte number with
         cpc    Delay2, temp    ;   0x000FA0
         cpc    Delay3, temp2   ;
         brcc   PC+2            ; less that FA0 so goes to high-speed
         rjmp   Start           ;
```

You will notice that instead of the expected line (**brcs   Start**) – i.e. branch to **Start** if the carry flag is set, we choose to skip the (**rjmp   Start**) line if the carry flag is clear. These two methods are clearly identical in their end result, but why introduce an extra line? The reason lies in the fact that the **brcs** can only branch to lines which are 64 instructions away. The **Start** line is, in fact, further away than this, and so must be branched to using the **rjmp** instruction. Points like this will be picked up when you try to assemble the program and are generally missed at the writing stage – so you don't have to start counting 60 odd lines whenever you introduce a **brcs** or similar instruction.

We then convert the time period of the oscillation into a frequency. To do this we need to take 4 000 000 and divide it by the length of time (in clock cycles) we have just measured. If we measured 40 000 clock cycles over one period, this will correspond to 100 Hz. There is a way to perform binary long division, but by far the simplest method of dividing **x** by **y** is to see how many times you can subtract **y** from **x**. This does take fewer instructions, but will take longer to run. We set up 4 000 000 = **0x3D0900**, spread over three temporary registers (**temp**, **temp2** and **temp3**). Every time we successfully subtract the number spread over **Delay1**, **Delay2** and **Delay3**, we increment a lower byte of the answer. When this overflows, we increment the higher byte. The answer will be between 1 and 1000 so we need only two bytes for the answer. The following lines set up the division:

```
ldi     temp, 0x00      ; moves 4 000 000 spread over 3
ldi     temp2, 0x09     ;   temporary registers
ldi     temp3, 0x3D     ;
clr     lowerbyte       ; resets the answer registers
clr     upperbyte       ;
```

EXERCISE 2.52   Write the *eight* lines of the loop called **Divide** which divides 4 000 000 by the number in the delay registers. (Hint: Call the next section **DoneDividing** and jump to this section when a subtraction was unsuccessful (carry flag was set).)

As with the high-speed section, we then convert the number in **lowerbyte** and **upperbyte** into hundreds, tens and ones. We can use the **DigitConvert** subroutine we have already written. The program then loops back to **LowSpeed**.

EXERCISE 2.53   What *two* lines wrap up the low-speed testing loop?

All that remains is the section called **TooSlow** which is branched to when the period of oscillations is more than one second. In this case we want to turn the displays off and send the AVR to sleep.

EXERCISE 2.54   Write the *three* lines which make up the **TooSlow** section.

You will have to remember to set up registers R0 to R11 with the correct seven segment code in the Init section. As you can use only the **ldi** instruction on registers R16–R31 you will have to move the numbers first into **temp**, and then move them into R0 to R11 using the **mov** instruction. Also, remember to set up PortD with one of the displays selected (e.g. 0b00000001), and define all your registers at the top of the program. It should now be ready for testing with the simulator. This may be worth building as it performs a useful function; however, you will notice that its resolution isn't great as you get only three-figure resolution between 100 Hz–999 Hz and 100 kHz–999 kHz. You may want to think about ways to improve the program to give three-figure resolution for all frequencies in the given range. In the coming chapters we will learn methods that will allow us to simplify this program hugely, and it will be worth coming back to this at the end and gleefully hack bits off to trim down the program.

Working on this larger program also introduces the importance of taking breaks. Even when you are 'in the zone' it is always a good idea to step back for a few minutes and relax. You will find you return looking at the bigger picture and may find you are overlooking something. Good planning and flowcharts help reduce such oversights. Another good piece of advice is to talk to people about decisions you have to make, or problems when you get stuck. Even if they don't know the first thing about microcontrollers, simply asking the question will surprisingly often reveal the answer.

3
Introducing the rest of the family

So far, we have been looking at the most basic types of AVR, the 1200 and the Tiny AVRs. I will now introduce some of the differences between these and other AVRs, so that in the subsequent chapters they might appear more familiar. Other models may benefit from extra memory called RAM. The allocation of memory differs in different models, but follows the arrangement shown in Figure 3.1.

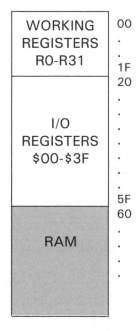

Figure 3.1

The first 32 addresses are the working registers and the next 64 are the I/O registers. So the key difference between those with RAM and those without is the presence of further memory spaces from $60 onwards. These can be accessed using the **ld** and **st** commands already introduced, and with other instructions now available on these more advanced models. A significant change to the working registers is the introduction of two more 2-byte register pairs. In addition to Z (made up of R30 and R31), there is now Y (made up of R28 and R29), and X (made up of R26 and R27). They can be used in any instruction that takes a Z (e.g. **ld**, **st**, **lpm** etc.).

Whilst there was a dedicated three-level stack on the 1200 and Tiny AVRs, the other models require that you tell them *where in the RAM you want your stack to be*. This means it is potentially as deep as the RAM address space, though obviously you may be wishing to give some of the RAM addresses a more glamorous purpose. What we will do is make the *last address of RAM* the top of the stack. In this way we have what looks like an upside-down stack, as shown in Figure 3.2, which works in exactly the same way as any other stack.

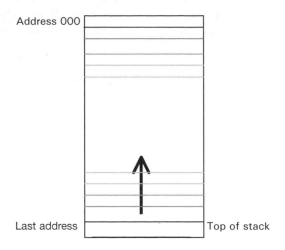

Address 000

Last address   Top of stack

Figure 3.2

The I/O registers **SPL** and **SPH** are the *stack pointer* registers (lower and higher bytes), and so we move into these the last address of the RAM. This is helpfully stored for us in the include file we read at the top of each program as RAMEND. We therefore load the lower byte of RAMEND into SPL and the upper byte into SPH, and thus point the stack to the end of the RAM. The instructions are:

```
ldi     temp, LOW(RAMEND)    ; stack pointer points to
out     SPL, temp            ;    last RAM address
ldi     temp, HIGH(RAMEND)   ;
out     SPH, temp            ;
```

This must take place in the **Init** section, before any subroutines are called. For devices with only 128 bytes of RAM, RAMEND is only 1 byte long, so the last two lines given above should be omitted.

Another major difference seen in the other models is a greater set of instructions. First, you are given greater flexibility with the **ld** and **st** instructions. You can make the 'long' registers X, Y or Z being used as an address pointer automatically increment or decrement with each load/store operation:

ld   reg, longreg+

This **ld**oads the memory location pointed to by a double register (i.e. X, Y or Z) into **reg**, and then adds one to longreg.

ld   reg, -longreg

This subtracts one from the double register (X, Y or Z), and then **ld**oads the memory location pointed to by that double register into **reg**. There are analogous instructions for **st**.

We can use this to shorten our multiple register clearing routine. In this case I have chosen to use X and the indirect address pointer, so this routine clears registers up to R25.

```
            clr   XL            ; clears XL
            clr   XH            ; clears XH
ClearLoop:  st    XH, X+        ; clears indirect address and increments X
            cpi   XL, 26        ; compares XL with 26
            brne  ClearLoop     ; branches to ClearLoop if ZL = 26
```

Other enhancements to load/store operations include:

ldd   reg, longreg+number

This **ld**oads the memory location pointed to by the Y or Z registers into **reg**, and then adds a **number** (0–63) to Y or Z. *(Note: doesn't work with X.)* There is an equivalent instruction for storing, **std**, which works in the same way. There is also a way to *directly address* memory in the RAM:

lds   reg, number

This **ld**oads the contents of memory at the address given by **number** into **reg**. The number can be between 0 and 65 535 (i.e. up to 64K). Similarly, **sts** stores the number in a register into a directly specified address.

Indirect jumps and calls are particularly useful and are specified by the number in the Z register:

icall   ; calls the address indirectly specified in Z
ijmp   ; jumps to the address indirectly specified in Z

Example 3.1   We have a program that has to perform one of five different
functions, depending on the number in a register called **Function**. By adding
Function to the current value of the program counter, and jumping to that
address, we can make the program branch out to different sections:

	clr	ZH	; makes sure higher byte is clear
	ldi	ZL, JumpTable	; points Z to top of table
	add	ZL, Function	; adds Function to Z
	ijmp		; indirectly jumps
JumpTable:	rjmp	Addition	; jumps to Addition section
	rjmp	Subtraction	; jumps to Subtraction section
	rjmp	Multiplication	; jumps to Multiplication section
	rjmp	Division	; jumps to Division section
	rjmp	Power	; jumps to Power section

Notice that **JumpTable** is loaded into Z, this is translated by the assembler as
the program memory address of the line it is labelling. We do this to initialize
Z to point to the top of the *branching table* (**rjmp   Addition**). Note that
loading **JumpTable** is equivalent to loading **PC+3**. The number in **Function** is
then added to Z, so that the number in **Function** (between 0 and 4) will make
the program jump to one of the five sections.

  You will no doubt remember the number of additions and subtractions we had
to do to 2-byte numbers in the frequency counter project. Here are two new
instructions that may help:

adiw	longreg, number
sbiw	longreg, number

These **add** or **sub**tract a **number** between 0 and 63 to/from one of the 16-bit
registers (X, Y or Z). The 'w' stands for **w**ord (16 bits). If there is an overflow
or carry this is automatically transferred onto the higher byte. Hence:

subi	XL, 50		sbiw	XL, 50
sbci	XH, 0	$\Longrightarrow$		

The two remaining instructions that are added to the repertoire of the more
advanced AVRs are:

push	register
pop	register

So far we have only been using the stack for the automatic storage of program
counter addresses when calling subroutines. Using these instructions, you can
push or **pop** the number in any working register on to or off of the stack.

Example 3.2 We can use the **push** and **pop** instructions to create a *palindrome detector*. A palindrome is essentially a symmetric sequence (like 'radar', 'dennis sinned' and 'evil olive'). We can massively simplify this problem by also requiring that we are given the length of the input sequence. We can use the length to find the middle of the input. We can also assume that the input is fed (as an ASCII character) into a register called **Input**. ASCII is a way to translate letters and symbols into a byte, so each letter corresponds to a particular byte-long number. So effectively we are looking for the sequence of bytes fed into **Input** to be palindromic (symmetric). We start by *pushing* the number in **Input** on to the stack. We do this for every new input until we reach the half-way point. We then start *popping* the stack and comparing it with the input. As long as each new input continues to be the same as the popped number, the sequence is potentially palindromic. If the new input fails to be the same as the popped number, we reject the input sequence. PinD, bit 0 will pulse high for 1 microsecond to indicate a new input symbol (we need this because we cannot just wait for the input symbol to change, as this would not respond to repeated letters).

First, we assume the length of the word is stored in **Length**. This has to be divided by two to get the half-way point. We will have to make a note if the length is odd or not. This is done by testing the carry flag; if it is high **Length** was odd and we shall set the T bit.

```
Start:    mov    HalfLength, Length  ; divides Length by 2 to get
          lsr    HalfLength          ;    HalfLength
          in     temp, SREG          ; copies Carry flag into T bit
          bst    temp, 0             ;    i.e. sets T-bit if Length is odd
```

Assuming the first input byte is in **Input**, we push it on to the stack and then wait for the pulse on PinD, bit 0. The pulse lasts 1 microsecond, so assuming a 4 MHz clock it must be tested at least once every four cycles. In the segment below, it is tested once every three cycles.

```
FirstHalf:  push   Input           ; pushes Input onto stack
Pulse:      sbis   PinD, 0         ; tests for pulse
            rjmp   Pulse           ; keeps looping
```

When a pulse is received (i.e. a new input symbol is ready), the program increments **Counter** which is keeping track of the input number. It compares this number with **HalfLength** and loops back as long as **Counter** is less than **HalfLength**.

```
          inc    Counter             ; counts the input number
          cp     Counter, HalfLength ; compares with half-way value
          brlo   FirstHalf           ; loops back to start and skips one
```

When **Counter** equals **HalfLength** we check the T bit to see if the length of the input is odd or even. If it is odd, we need to ignore the middle letter, so we reset the T bit and loop back to **Pulse** which will wait until the next input is ready. If the length is even we can skip on to test the second half of the input.

```
brtc    SecondHalf      ; test T bit
clt                     ; clears T bit
rjmp    Pulse           ;
```

We have now passed the half-way point in the sequence and now the new input symbols must match the previous ones. The top of stack is popped and compared with the current input. If they are not equal the sequence is rejected.

```
SecondHalf: pop     Input2          ; pops stack into Input2
            cp      Input, Input2   ; compares Input and Input2
            brne    Reject          ; if different reject sequence
```

As before, we then increment **Counter** and test to see if **Counter** equals **Length**. If it does, the testing is over and we can accept the input. If we haven't yet reached the end the program then waits for the input to change, and then loops back to **SecondHalf**.

```
        inc     Counter         ; counts the input number
        cp      Counter, Length ; compares with total length
        breq    Accept          ; end of sequence so accept

Pulse2: sbis    PinD, 0         ; waits for pulse
        rjmp    Pulse2          ;
        rjmp    SecondHalf      ; loops back when new input is
                                ;   ready
```

You might want to play around with this on the simulator, but don't forget to set up the top of the stack as described at the start of the chapter. You may also want to think about how to remove the need to be given the length of the input sequence. If you want to find out more about this, you may want to find a book on *Formal Languages* and *Parsing*.

4
Intermediate operations

Interrupts

So far we have always had to test for certain events ourselves (e.g. test for a button to be pressed, test if T/C0 has overflowed etc.). Fortunately there are a number of events which can automatically alert us when they occur. They will, if correctly set up, interrupt the normal running of the program and jump to a specific part of the program. These events are called *interrupts*.

On the 1200, the following interrupts are available:

- Interrupt when the INT0 pin (PD2) is low
- Interrupt when there is a rising edge on INT0
- Interrupt when there is a falling edge on INT0
- Interrupt when T/C0 overflows
- Interrupt when the Analogue Comparator triggers a result

The first three constitute an *external interrupt* on INT0, and are mutually exclusive (i.e. you can enable only one of the three interrupts at any one time). The significance of the *Analogue Comparator* will be discussed later on in the chapter. When an interrupt occurs, the program will jump to one of the addresses at the start of the program. These addresses are given by what is known as the *interrupt vector table*. The interrupt vector table for the 1200 is shown in Table 4.1, the tables for the other AVR types are shown in Appendix E.

Table 4.1

Type of Interrupt/Reset	Program jumps to address ...
Power-on/Reset	0x000
External interrupt on INT0	0x001
T/C0 overflow interrupt	0x002
Analogue comparator interrupt	0x003

For example, when the T/C0 overflow interrupt is enabled, and T/C0 overflows, the program drops what it's doing and jumps to address 0x002 in the program memory. When using all three interrupts, the start of the program should look something like the following:

```
rjmp    Init            ; first line executed
rjmp    ExtInt          ; handles external interrupt
rjmp    OverflowInt     ; handles TCNT0 interrupt
rjmp    ACInt           ; handles A. C. interrupt
```

This will ensure the program branches to the correct section when a particular interrupt occurs (we will call these *interrupt handling routines*). We can enable individual interrupts using various registers. The enable bit for the External INT0 interrupt is bit 6 in an I/O register called **GIMSK** (General Interrupt Mask). Setting this bit enables the interrupt, clearing it disables it. The enable bit for the TCNT0 overflow bit is bit 1 in the **TIMSK** I/O register (Timer Interrupt Mask). However, all of these interrupts are overridden by an interrupts 'master enable'. This is a master switch which will disable all interrupts when off, and when on it enables all *individually enabled* interrupts. This bit is the **I** bit in **SREG** (you may want to glance back to page 73).

The External INT0 interrupt can be set to trigger in one of three different circumstances, depending on the states of bits 0 and 1 of the **MCUCR** I/O register (the one that also holds the sleep settings). This relation is shown in Table 4.2.

Table 4.2

MCUCR Bit1	Bit 0	Interrupt occurs when ...
0	0	INT0 is low
0	1	*Invalid selection*
1	0	There is a falling edge on INT0
1	1	There is a rising edge on INT0

When an interrupt occurs, the value of the program counter is stored in the stack as with subroutines, so that the program can return to where it was when the interrupt handling is over. Furthermore, when the interrupt occurs, the master interrupt enable bit is *automatically cleared*. This is so that you don't have interrupts occurring inside the interrupt handling routine which would then lead to a mess of recursion. You will probably want to re-enable the master interrupt bit upon returning from the interrupt handling routine. Fortunately there is a purpose-built instruction:

```
reti                    ;
```

This **ret**urns from a subroutine and at the same time enables the master interrupt bit.

Each interrupt also has an *interrupt flag*. This is a flag (bit) that goes high when an interrupt *should* occur, even if the global interrupts have been disabled and the appropriate interrupt service routine isn't called. If the global interrupts are disabled (for example, we are already in a different interrupt service routine) you can test the flag to see if any interrupts have occurred. Note that these flags stay high until reset, and an interrupt service routine will be called if the flag is high *and* the global interrupt bit is enabled. So you must reset all flags before enabling the global interrupt bit, just in case you have some interrupt flags lingering high from an event that occurred previously. Interrupt flags are reset by *setting* the appropriate bit – this sounds counterintuitive but it's just the way things are! The T/C0 Overflow interrupt flag is found in bit 1 of **TIFR** (**T**imer **I**nterrupt **F**lag **R**egister – I/O number $38), and the INT0 interrupt flag is in bit 6 of **GIFR** (**G**eneral **I**nterrupt **F**lag **R**egister – I/O number $3A).

Program K: reaction tester

- Interrupts
- Random number generation
- Seven segment displays

The next example program will be a reaction tester. A ready button is pressed, then an LED will turn on a random time later (roughly between 4 and 12 seconds). The user has to press a button when they see the LED turn on. The program will measure the reaction time of the user and display it in milliseconds on three seven segment displays. If the user presses the button before the LED turns on they will be caught cheating. The circuit diagram for the project is shown in Figure 4.1, and the flowchart in Figure 4.2.

We will be using the External INT0 and TCNT0 Overflow interrupts, so you will have to make the necessary changes to the top of the program. Note that as we will not be using the Analogue Comparator interrupt we don't need any particular instruction at address 0x003.

EXERCISE 4.1   What are the first *three* instructions of the program?

Write the Init section, setting T/C0 to count internally at CK/1024. You will have to enable the External INT0 and T/C0 Overflow interrupts, but don't set the master enable just yet. Set the External INT0 interrupt to occur when INT0 is low (i.e. when the button is pressed).

EXERCISE 4.2   What are the *six* lines which individually enable the interrupts?

At **Start** we first call the **Display** subroutine, and then test the 'Ready' button (PinD, 1). Keep looping until the Ready button is pressed.

EXERCISE 4.3   What *three* lines achieve this?

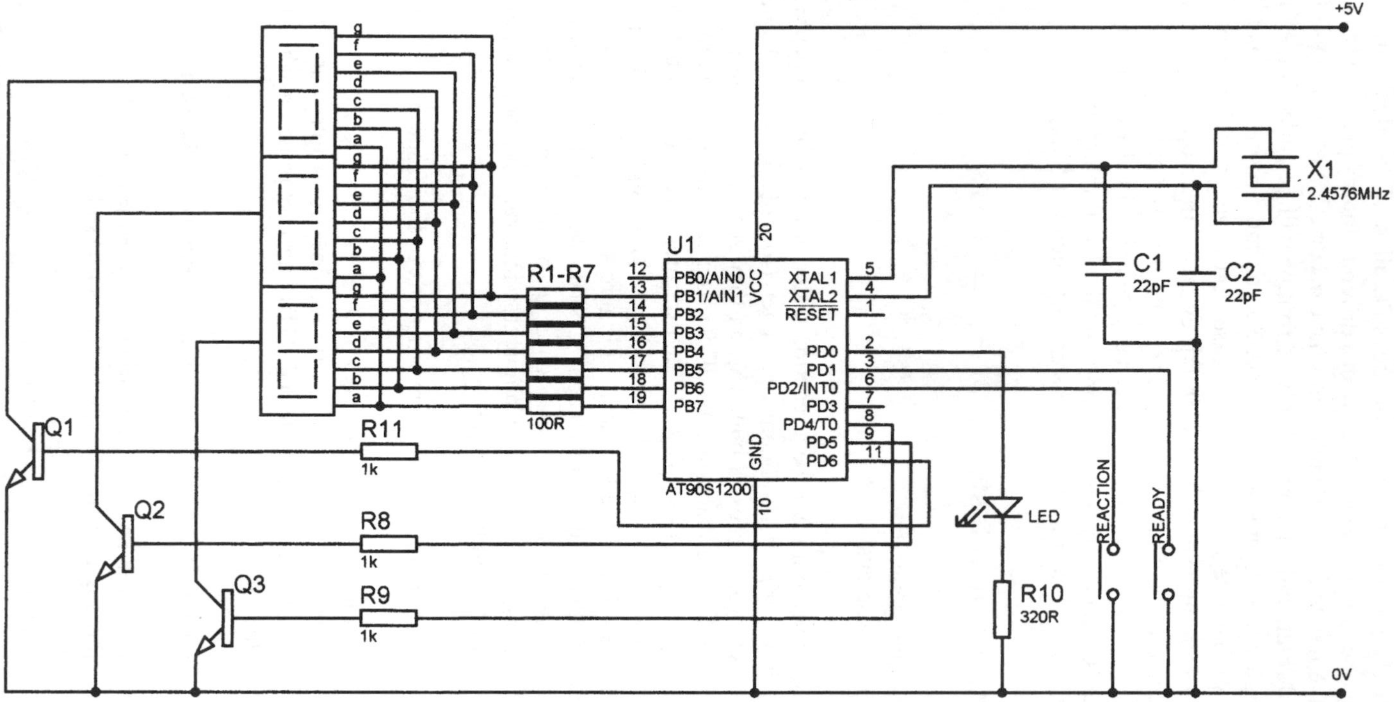
+5V
X1
2.4576MHz
C1
22pF
C2
22pF
U1
PB0/AIN0
PB1/AIN1
PB2
PB3
PB4
PB5
PB6
PB7
VCC
GND
XTAL1
XTAL2
RESET
PD0
PD1
PD2/INT0
PD3
PD4/T0
PD5
PD6
AT90S1200
20
10
12
13
14
15
16
17
18
19
5
4
1
2
3
6
7
8
9
11
R1-R7
100R
g
f
e
d
c
b
a
R11
1k
R8
1k
R9
1k
Q1
Q2
Q3
LED
R10
320R
REACTION
READY
0V

Figure 4.1

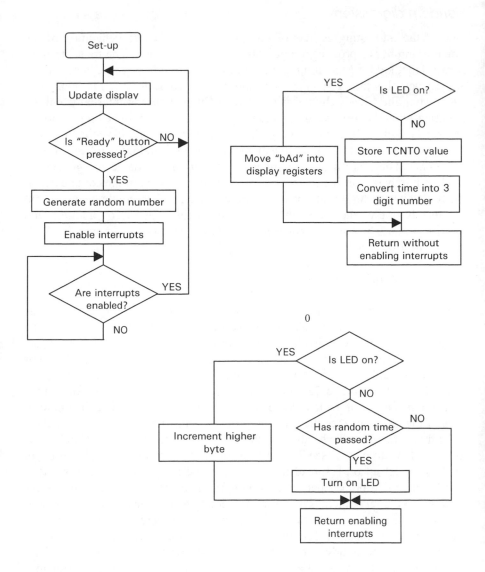

Figure 4.2

The **Display** subroutine will be almost exactly like the one in the frequency counter project. The only difference lies in the selection of the correct display. Instead of rotating between bit 0 and bit 2 of Port D, this part of the subroutine will have to rotate between bit 4 and bit 6, testing bit 7 to see when it has gone too far. Make the necessary changes to the subroutine and copy it in. We now need to create a random time delay.

Random digression

One of the interesting aspects of this program will be the generation of the random number to produce a time delay of random length. The most straight-forward method for generating random numbers is to rely on some human input and convert this into a number. For example, we could look at the number in T/C0 when the 'Ready' button is pressed. T/C0, if counting internally, will be counting up and overflowing continuously, and so its value when the button is pressed is likely to be random. Very often, however, we don't have the luxury of a human input, and so we have to generate a string of random numbers. How is this done? There are a large number of algorithms available for generating random numbers, varying in complexity. We are restricted in the complexity of the functions we can straightforwardly apply using AVR assembly language, but fortunately one of the more simple algorithms relies purely on addition and multiplication. The *Linear Congruential Method* developed by Lehmer in 1948[1] has the following form:

$$I_{n+1} = \mathrm{mod}_m(aI_n + c)$$

This generates the next number in the sequence by multiplying the previous number by a, adding c, and taking the result modulo m. $\mathrm{mod}_m(x)$ is equal to the remainder left when you divide x by m. Conveniently, the result of every operation performed in an AVR program is effectively given in modulo 256. For example, we add 20 to 250. The 'real' answer is 270; however, the result given is 14. 14 *is* '270 modulo 256' or $\mathrm{mod}_{256}(270)$. There are a number of restrictions on the choice of a and c in the above equation that maximize the randomness of the sequence (see the reference for more info). Given that the quickest algorithm is that with the smallest multiplier (a), we will choose $a = 5$ and $c = 1$. You also have to pick a 'seed' – the first number in the sequence (I_0). You can set this model up on a spreadsheet and examine its *quasirandom* properties. First, you should notice that the randomness of the sequence does not appear sensitive to the seed; there is therefore no need to pick a particular one. You will also notice the sequence repeats itself every 256 numbers – this is an unfortunate property of the algorithm. Picking a larger modulus will increase the repetition period accordingly. We could use modulo 65 536 by using one of the 2-byte registers (X, Y or Z) and the **adiw** instruction. This would result in a sequence that repeats only every 65 536 numbers! For our purposes with the reaction tester, a period of 256 is quite acceptable.

To convert this random number into a random time we do the following. The maximum time is 10 seconds, and the T/C0 will overflow every 256 counts = 256/2400 = 0.066 second. We therefore would like a counter with a value roughly 61 and 183. You might notice the difference between these numbers is

---

[1] See reference on random numbers in Appendix I.

not far off 128 (it is in fact 122). Our life is made a lot easier if the difference is 128, so as the times needed are quoted only as approximate figures, we can use a counter that goes from 60 to 188 which will perform adequately. To convert our random number between 0 and 255 we first divide by two, then add 60.

Returning to the program, we will use register **Random** to hold the random number. We need then to multiply this by five (add it to itself four times), and then add one to it.

EXERCISE 4.4   What *six* lines will generate the next random number?

EXERCISE 4.5   What *three* lines will copy **Random** into **CountX**, divide **CountX** by two, and then add 60.

We then need to reset the higher byte of the timer (**TimeH**), turn off the displays (clear **PortB**), reset all the interrupt flags, and then set the master interrupt enable.

EXERCISE 4.6   Which *six* lines will reset **TimeH**, **PortB** and the interrupt flags?

There is a particular instruction for setting the master interrupt enable:

      sei   ; Sets the interrupt enable bit.

The rest of the program is a loop which just tests the interrupt enable bit, and loops back to **Start** when it has been cleared. This is because after an External INT0 interrupt, the master interrupt bit will *not* re-enable interrupts and upon returning the program will loop back to **Start**. In contrast, after a T/C0 related interrupt the interrupts *will* be re-enabled so the program will stay in the loop.

EXERCISE 4.7   What *three* lines finish off the main body of the program?

Looking first at the T/C0 overflow interrupt handling routine (**TInt**), we see that the first test is to see whether or not the LED (PinD, 0) is on. If it is off we should be timing out the random time to see when to turn it on. If it is already on we should be incrementing the higher byte of our timing registers (**TimeH**). If the time exceeds the maximum that can be displayed on the scope, we should move '-HI' into the display registers and return without enabling interrupts.

The T/C0 is counting up 2400 times a second (with a register counting the higher byte as well). We need to convert this to milliseconds (i.e. something counting 1000 times a second). To do this we can multiply the 2-byte number by 5 and then divide by 12. Applying the reverse procedure to 999 (the maximum response time) we get 2397 = 95D. It would be much easier if we were testing only to see if the higher byte had reached a certain value (e.g. A00). This is easy to do by resetting T/C0 to 0xA2 when the LED is turned on, and then subtract the 0xA2 back off the final answer at the end of the day:

```
TInt:        sbic    PinD, 0         ; tests LED
             rjmp    TInt_LEDon      ; jumps to different section if on

             dec     CountX          ; decrements random counter
             breq    PC+2            ; skips if clear
             reti                    ; returns otherwise

             sbi     PortD, 0        ; turns on LED when time passes
             ldi     temp, 0xA2      ; initializes TCNT0 to 0xA2
             out     TCNT0, temp     ; to facilitate testing for max
             reti                    ;

TInt_LEDon:
             inc     TimeH           ; increments higher byte
             cpi     TimeH, 0x0A     ; tests for maximum time
             breq    PC+2            ; skips if the user is too slow
             reti                    ;
             ldi     Hundreds, 13    ; -
             ldi     Tens, 14        ; H
             ldi     Ones, 1         ; I
             ret                     ; returns without setting I-bit
```

The External INT0 interrupt handling routine is more straightforward – we will call it **ExtInt**. This also involves testing the LED first. If it isn't on this means the user has cheated by pressing the button before the LED has turned on. In this case, we move numbers 10, 11 and 12 into **Hundreds**, **Tens** and **Ones** respectively in order to display 'bAd', and then return *without* re-enabling the master interrupt bit. If the LED is on, the press is valid, and so we have to halt the T/C0 and store the current time by moving T/C0 into **TimeL**. It is possible, however unlikely, that the T/C0 overflowed just after the INT0 interrupt occurred. We therefore need to test the T/C0 overflow interrupt flag, and increment **TimeH** if it is set. Then the total reaction time (split up over **TimeL** and **TimeH**) needs to have 0xA2 subtracted from it (as this was artificially added). It must then be multiplied by 5 and divided by 12.

EXERCISE 4.8   Which *12* lines test the LED at the start of **ExtInt**, test the LED, jump to a section called **Cheat** if it isn't on, and halt the T/C0 and store the current value, incrementing **TimeH** if necessary? 0xA2 should then be subtracted from the total reaction time, and T/C0 should be restarted at CK/1024.

EXERCISE 4.9   Which *four* lines form the **Cheat** section?

After subtracting 0xA2 we need to multiply the time by 5. As the time is split over two registers we need to use the **adc** to add a carry to the higher byte if and when there is a carry:

	ldi	Count4, 4	; loads a counter with 4
	mov	temp, TimeL	; stores time in temp and tempH
	mov	tempH, TimeH	;
Times5:	add	temp, TimeL	; adds TimeL to itself
	adc	tempH, TimeH	; adds TimeH and Carry to itself
	dec	Count4	; does this 4 times
	brne	Times5	;

The product is now held over **temp** and **tempH**. We then divide the result by 12. The simplest way to do this is to see how many times we can subtract 12 from the total.

EXERCISE 4.10   *Challenge!* What *nine* lines will first clear **TimeL** and **TimeH**, and then enter a loop which divides the 2-byte number stored between **temp** and **tempH** by 12, leaving the result in **TimeL** and **TimeH**. (To skip out of the loop jump to the **DigitConvert** section.)

DigitConvert converts the 2-byte number into a three-digit number (this is copied from the frequency counter with the register names changed accordingly). Instead of the **ret** instruction at the end of the section, write **rjmp   Start**.

You will have to set up all the registers (R0–R14) that hold the seven segment codes in the Init section. Registers R10, R11, R12, R13 and R14 hold the codes for a 'b', 'A', 'd', '-' and 'H' respectively. You can double check you've done everything correctly by looking at Program K in Appendix J. It should be quite fun to try this one out. Of course, the simplest way of using an AVR as a reaction tester is to get a friend to hold it between your fingers and drop it, and then see how far down the chip you caught it!

Analogue comparator

Another useful feature on most of the AVRs is an analogue comparator (AC) which compares the voltages on two pins (called **AIN0** and **AIN1** = PB0 and PB1 on the 1200) and changes the state of a bit depending on which voltage is greater. This is all controlled by the **ACSR** I/O register, whose bit assignments are shown in Figure 4.3.

Bit 7 is simply an on/off switch for the AC. You should disable the AC interrupt (clear bit 3) before disabling the AC otherwise an interrupt might occur when you try to switch it off. Bits 0 and 1 dictate what triggers an AC interrupt in terms of the AC result (i.e. interrupt when the AC result changes, when it rises, or when it falls). The remaining bits are self-explanatory.

ACSR – **A**nalogue **C**omparator **C**ontrol and **S**tatus **R**egister

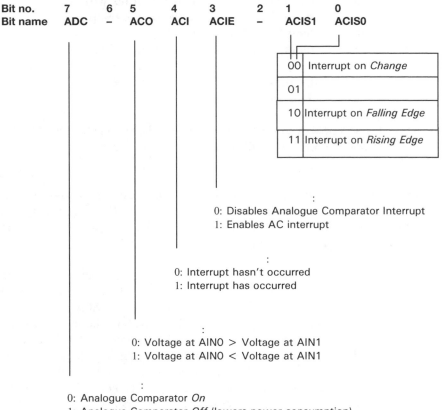

Figure 4.3

Program L: 4-bit analogue to digital converter

● Analogue comparator

This next project is very much a case of doing what you can with what you have. Some of the more advanced AVRs have full-blown 10-bit analogue to digital converters, and so with these the ability to create a 4-bit converter is clearly of limited value. However, many AVRs don't benefit from this luxury, being blessed with only a comparator, and in these cases the following program can be useful. The key to this project is using a summing amplifier to create one of 16 possible reference voltages. By running through these reference voltages and comparing them with the input signal, we can determine the input voltage with 4-bit resolution and within four cycles of the loop. The circuit diagram is shown in Figure 4.4, pay particular attention to how the summing amplifier works. For more information on summing amplifiers, see the reference[2]. The straightforward flowchart is shown in Figure 4.5.

PD0 to PD3 control which reference voltage is being fed to the comparator, as summarized in Table 4.3.

Table 4.3

0000	0 V	1000	2.5 V
0001	0.312 V	1001	2.812 V
0010	0.625 V	1010	3.125 V
0011	0.937 V	1011	3.437 V
0100	1.25 V	1100	3.75 V
0101	1.562 V	1101	4.062 V
0110	1.875 V	1110	4.375 V
0111	2.187 V	1111	4.687 V

Write the Init section, remembering to turn on the analogue comparator by setting bit 7 of **ACSR**. Leave the AC interrupt off. At **Start** we first set up PortD with 0b00001000. This sets the most significant bit of the voltage selector and thus feeds 2.5 V into AIN0. This is then compared with the input at AIN1. If the input is higher than the reference, bit 5 of **ACSR** will be high, otherwise bit 5 will be low. If the input is higher than the reference, the answer is greater than 1000 and so we leave bit 3 of the reference high and set bit 2. If the input is lower than the reference, the answer is less than 1000 and so we clear bit 3, and then set bit 2.

EXERCISE 4.11   Write the *five* lines which set up PortD with the initial value and then test the AC result. If the AC result is low, clear bit 3 of PortD. In either case set bit 2 of PortD.

EXERCISE 4.12   Repeat the above for the remaining bits (*eight* more lines).

EXERCISE 4.13   *Challenge!* Write the *four* lines that transfer the resulting state of PD0-3 to the output bits (PB4-7), and then loop back to **Start**.

---

[2] See references: *Introducing Electronic Systems*, M. W. Brimicombe (1997) Nelson Thornes.

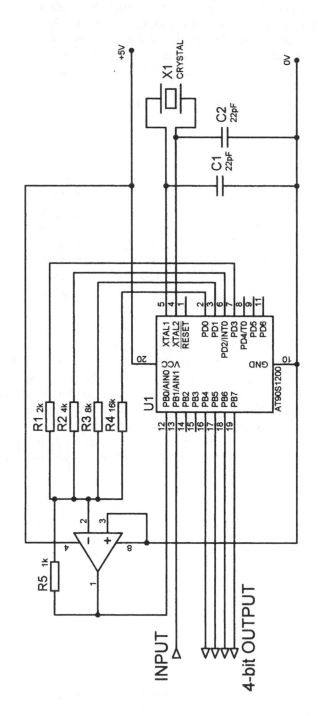

Figure 4.4

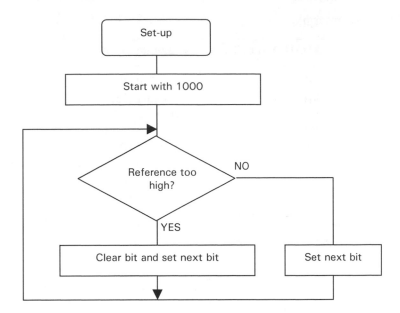

Figure 4.5

10-bit analogue to digital conversion (ADC)

Other AVR models such as the Tiny15, 4433 and 8535 have a built-in 10-bit A/D converter. This works in much the same way as the 4-bit converter we built in the previous section, except it is all done for us automatically and internally. The voltage on one of the analogue input channels is measured (with respect to the voltage on a reference pin AREF), converted into a 10-bit binary number, and stored over two I/O registers called **ADCL** and **ADCH** (which stand for **ADC** Result **L**ower byte and **ADC** Result **H**igher byte). There are two basic modes of operation: *Free Running* and *Single Conversion*. In 'Free Running' the ADC repeatedly measures the input signal and constantly updates **ADCL** and **ADCH**. In 'Single Conversion' the user must initiate every AD conversion themselves.

For the 4433 and 8535, the pin being read is selected using the I/O register called **ADMUX ($07)**. The bit assignment is shown in Table 4.4, all other bits are not used.

If you want to test a number of channels, you can change the **ADMUX** register, and the channel will be changed immediately, or, if an AD conversion is in progress, after the conversion completes. This means you can scan through channels in 'Free Running' mode more easily, as you can change the channel during one conversion, and the next conversion will be on the new channel.

The rest of the ADC settings are held in the **ADCSR** (**ADC S**tatus **R**egister), I/O register **$06**. The bit assignments are shown in Figure 4.6.

Table 4.4

ADMUX bits 2,1,0	Analogue input
000	Channel 0 (PA0)
001	Channel 1 (PA1)
010	Channel 2 (PA2)
011	Channel 3 (PA3)
100	Channel 4 (PA4)
101	Channel 5 (PA5)
110	Channel 6 (PA6)
111	Channel 7 (PA7)

ADCSR – ADC Status Register ($06)

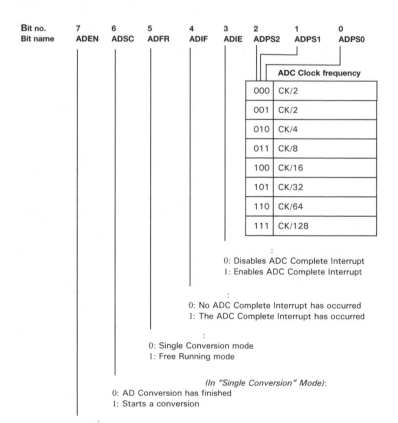

Bit no.	7	6	5	4	3	2	1	0
Bit name	ADEN	ADSC	ADFR	ADIF	ADIE	ADPS2	ADPS1	ADPS0

	ADC Clock frequency
000	CK/2
001	CK/2
010	CK/4
011	CK/8
100	CK/16
101	CK/32
110	CK/64
111	CK/128

0: Disables ADC Complete Interrupt
1: Enables ADC Complete Interrupt

0: No ADC Complete Interrupt has occurred
1: The ADC Complete Interrupt has occurred

0: Single Conversion mode
1: Free Running mode

(In "Single Conversion" Mode):
0: AD Conversion has finished
1: Starts a conversion

0: ADC *Off* (lowers power consumption)
1: ADC *On*

Figure 4.6

Bits 0 to 2 control the frequency of the ADC clock. This controls how long each conversion takes and also the accuracy of the conversion. A clock between 50 kHz and 200 kHz is recommended for full, 10-bit, accuracy. Frequencies above 200 kHz can be chosen if speed of conversion is more important than accuracy. For example, a frequency of 1 MHz gives 8-bit resolution, and 2 MHz gives 6-bit resolution. The ADC complete interrupt occurs (if enabled) when an ADC conversion has finished. The other bits are straightforward.

The ADC on the Tiny15 is slightly more advanced, offering features such as internal reference voltages and differential conversion (i.e. measuring the voltage difference between two pins). Moreover, in the case of the 4433 and 8535 the 10-bit ADC result is stored with the lower byte in **ADCL** and the remaining two msb's in **ADCH**. In the case of the Tiny15, you have the choice between this arrangement, and storing the upper byte in **ADCH** and the remaining two lsb's in **ADCL**. These changes all take place in the **ADMUX** register, shown in Figure 4.7.

Looking at bits 0 to 2 again, we see the option to look at the voltage difference between pins, namely ADC2 (PB3) and ADC3 (PB4). These inputs are put through a differential amplifier, and then measured using the ADC. The differential amplifier can either have a gain of x1 or x20. You will notice that two of the settings give the difference between ADC2 and itself! This is used for calibration purposes, as the differential amplifier used in the difference setting will have a small offset. By measuring this offset and subtracting from the answer of your real difference measurement, you will improve the accuracy of your result.

Another handy feature if you are interested in a high accuracy conversion is to send the chip to sleep and perform an AD conversion whilst in sleep. This helps eliminate noise from the CPU (central processing unit) of the chip. An ADC complete interrupt can then be used to wake up the chip from sleep. This method is demonstrated in Example 4.1.

Example 4.1

```
ldi     temp, 0b10001011   ; enables ADC, Single Conversion
out     ADCSR, temp        ; enables AD Complete Interrupt
ldi     temp, 0b00101000   ; enables sleep,
out     MCUCR              ; 'AD Low Noise mode'
sleep                      ; goes to sleep – this automatically
                           ;    starts AD Conversion
```

When the AD conversion completes, the AD conversion interrupt routine will be called (address $008 on the Tiny15, and address $00E on the 4433 or 8535), when the program returns from the routine it will carry on from the line after the sleep instruction.

ADMUX – ADC Multiplexer ($07)

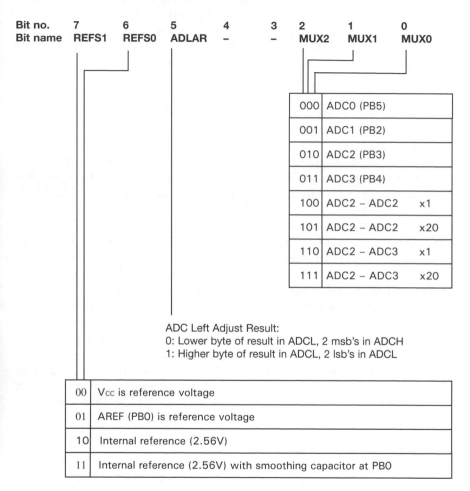

Bit no.	7	6	5	4	3	2	1	0
Bit name	REFS1	REFS0	ADLAR	–	–	MUX2	MUX1	MUX0

000	ADC0 (PB5)
001	ADC1 (PB2)
010	ADC2 (PB3)
011	ADC3 (PB4)
100	ADC2 – ADC2 x1
101	ADC2 – ADC2 x20
110	ADC2 – ADC3 x1
111	ADC2 – ADC3 x20

ADC Left Adjust Result:
0: Lower byte of result in ADCL, 2 msb's in ADCH
1: Higher byte of result in ADCL, 2 lsb's in ADCL

00	Vcc is reference voltage
01	AREF (PB0) is reference voltage
10	Internal reference (2.56V)
11	Internal reference (2.56V) with smoothing capacitor at PB0

Figure 4.7

Program M: voltage inverter

● Analogue to digital conversion
● Digital to analogue conversion

We can use ADCs to make digital to analogue converters. The trick to this is to use the output to charge up a capacitor until it reaches the required output voltage. The AVR's output then goes open circuit (turns itself into an input). The capacitor will then slowly discharge through the input impedance of whatever is reading it, lowering the analogue output. Meanwhile another input is monitoring the voltage of the analogue output. If it falls below a certain mark, the AVR's output is turned on again to top up the analogue output. To lower the analogue voltage, the AVR output is cleared to 0 to discharge the capacitor quickly. Figure 4.8 illustrates this technique, though the jaggedness of the final output is exaggerated.

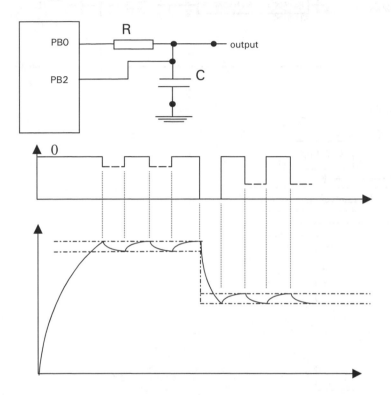

Figure 4.8

R should be made small enough to allow quick response time, and C high enough to give a smooth output. We will demonstrate this with a project that takes an input, i, between 0 and 5 V, and outputs $(5 - i)$. For example, 2 V in becomes 3 V out. The circuit diagram is shown in Figure 4.9, and the flowchart in Figure 4.10.

In the Init section, we will have to enable A/D conversion, and select ADC0 to start with. We would like maximum accuracy, and so require a clock speed that is less than 200 kHz. We will be using the internal oscillator which runs at 1.6 MHz. This means that an ADC clock of CK/8 (200 kHz) will be acceptable. The ADC should be single mode, and set the 'Left Adjust' bit so that the upper byte of the ADC result is in ADCH and the two lsbs in ADCL. Finally, let V_{CC} be the reference voltage, and start an AD conversion.

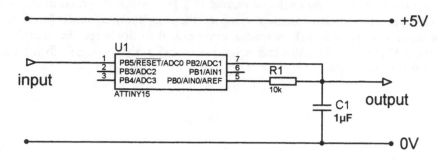

Figure 4.9

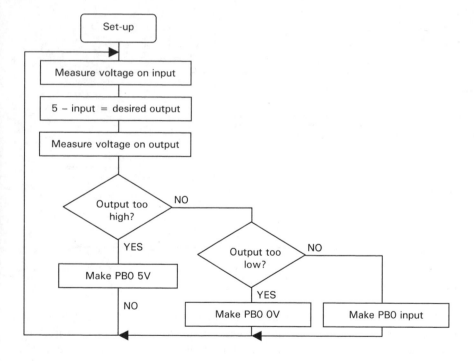

Figure 4.10

EXERCISE 4.14   What numbers should be moved into **ADCSR** and **ADMUX** in the Init section?

Write the whole of the Init section. Initially make PB0 an output. PB5 and PB2 should be inputs. Once the AVR reaches Start, the ADC0 channel should be selected (by clearing ADMUX, bit 0), and an A/D conversion should be started (by setting ADCSR, bit 6). When the A/D conversion is over, this bit will be cleared, so we can test this bit and wait for it to set.

EXERCISE 4.15   What *four* instructions start an A/D conversion on ADC0 and wait for it to complete?

Once the conversion is complete, the input voltage will be stored in registers ADCL and ADCH. There is no need for the full 10-bit accuracy, and so we will simply use 8 bits. With Left Adjust enabled, this simply involves reading the number from ADCH. To perform the function (5 – input voltage) we simply invert the result (ones become zeros and vice versa). Invert the results using the **com** instruction, and store the result in a working register called **Desired** (this represents the voltage we *want* on the output).

EXERCISE 4.16   Which *six* instructions store and invert the measurement of the input voltage, change the input channel to select ADC1, and start a new conversion? It should also wait in a loop until the current conversion finishes.

Now the voltage on the output has been read and can be compared with the desired voltage. Save the measured voltage from ADCH into a working register called **Actual** (the *actual* voltage on the output). Then use the compare (**cp**) and branch-if-lower (**brlo**) instructions to jump to sections called **TooHigh** (the actual output is higher than the desired output), or **TooLow** (the actual output is less than the desired output).

EXERCISE 4.17   Which *seven* lines perform these tests and branch out as required? If the actual and desired voltages are equal, PB0 should be made an input (by clearing DDRB, bit 0) and then the program should jump back to **Start**.

The **TooHigh** section needs to lower the output, and so PB0 is made an output (by setting DDRB, bit 0) and then made low (0V) to discharge the capacitor and lower the output. **TooLow** needs to raise the output, and so PB0 is made an output and made high (5V) to charge up the capacitor.

EXERCISE 4.18   Write the *six* lines that make up the **TooHigh** and **TooLow** sections. The end of both sections should jump back to **Start**.

That wraps up Program M. You may want to experiment a little and make the device perform more complicated functions on the input, or perhaps on *two* inputs. Perhaps you can make some form of audio mixer by summing two input channels, or subtract the left and right channels of an audio signal to get a 'pseudo-surround sound' output. As you can see, there are a number of interesting projects that can be based around the above, and all on the little Tiny15 chip!

EEPROM

In addition to the RAM and program memory that we have already seen, many AVRs have an additional memory store which combines the flexibility of RAM, with the permanence of program memory. Unlike the RAM, the EEPROM will keep its values when power is removed and unlike the program memory, the EEPROM can be read and written to while the program runs. EEPROM stands for **E**lectrically **E**rasable **R**ead-**O**nly **M**emory. There are three I/O registers associated with the EEPROM:

- **EEAR** – The **r**egister which holds the **a**ddress being written to/read from the **EE**PROM
- **EEDR** – The **r**egister which holds the **d**ata to be written to/read from the **EE**PROM
- **EECR** – The **r**egister which holds **c**ontrols the **EEPROM**
  - Set bit 0 of EECR to read from the EEPROM
  - Set bit 1 of EECR to write to the EEPROM

The 1200 has 64 bytes of EEPROM, though other AVRs can have much more (up to 512 bytes). The write operation takes a certain amount of time. To wait until the writing process is over, test bit 1 of **EECR** (the one you set to start the write) – when the writing finishes the bit is cleared automatically.

Example 4.2   To write the number 45 to EEPROM address 0x30, we would write the following:

```
          ldi    temp, 0x30      ; sets up address to write to
          out    EEAR, temp      ;
          ldi    temp, 45        ; sets up data to write
          out    EEDR, temp      ;
          sbi    EECR, 1         ; initiates write
EEWait:   sbic   EECR, 1         ; waits for write to finish
          rjmp   EEWait          ; loops until EECR, 1 is cleared
```

Example 4.3   To read address 0x14 of the EEPROM we write the following. At the end of the segment of code, the data held in address 0x14 will be in **EEDR**.

```
          ldi    temp, 0x14      ; sets up address to read
          out    EEAR, temp      ;
          sbi    EECR, 0         ; initiates read
                                 ; data now held in EEDR
```

EXERCISE 4.19 *Challenge!* Write a routine which sets up addresses 0x00 to 0x0F of the EEPROM to be an ASCII look-up table. This means address 'n' of the EEPROM holds the ASCII code for the 'n' character (i.e. the code for numbers 0–9, A, B, C, D, E and F). The ASCII codes for the relevant characters are given in Appendix G. The routine should be *14* lines long.

There are two ways to program the EEPROM when you are programming your chip. In AVR Studio, you can go to View → New Memory View (Alt + 4) and select EEPROM. This will give you a window with EEPROM memory locations. Simply type in the values you wish to program into the EEPROM, and when you select the programmer (e.g. STK500), select 'Program EEPROM' and choose 'Current Simulator/Emulator Memory'. This will load the contents of the EEPROM window onto the EEPROM of the chip. An easier way is to specify what you want to write to the EEPROM in your program itself. Use the **.eseg** directive (**E**EPROM **seg**ment) to define EEPROM memory. What you write after that will be written to the EEPROM. If you want to write normal code after this, you must write **.cseg** (**c**ode **seg**ment).

Example 4.4

```
.eseg                  ; writes what follows to the EEPROM
.db    0x04, 0x06, 0x07 ;
.db    0x50            ;

.cseg                  ; writes what follows to the program
                       ;   memory
ldi    temp, 45        ;
```

The **.db** directive stores the byte(s) which follow to memory. This particular code writes 0x04, 0x06, 0x07 and 0x50 to memory locations 00–03 in the EEPROM. Note that this is *not* a way to change the EEPROM during the running of the programming – it is only a way to tell the programmer what to write to the EEPROM *when you are programming the chip*. Directives such as **.org** can be used to select specific addresses in the EEPROM. On the 1200, which doesn't support the **lpm** instruction, it is a better use of resources to store the seven segment look-up table in the EEPROM, than in registers R0–R10, as previously done.

16-bit timer/counter 1

Some AVRs, such as the 2313, have a separate 16-bit timer/counter in addition to the 8-bit TCNT0. This is called Timer/Counter 1, and is quite useful as the need for markers and counters to time natural time lengths becomes greatly reduced. The number in Timer/Counter 1 (T/C1) is spread over two I/O registers: **TCNT1H** (higher byte) and **TCNT1L** (lower byte). The T/C1 can be prescaled separately to T/C0 (i.e. it can be made to count at a different speed), and can also be made a counter of signals on its own input pin: T1 (as opposed to T0 which is the T/C0 counting pin). If the T/C1 is counting up at 2400 Hz, the 16 bits allow us to time up to 27 seconds without the need for any further counters. One very important point to note with this 2-byte timer/counter is that when you read T/C1, the 2 bytes must be read *at the same time*, otherwise there is a chance that in between storing the lower and higher bytes, the lower byte overflows, incrementing the higher byte, which lead to a large error in the stored answer. In order to do this you must therefore *read the lower byte first*. When you read in the **TCNT1L**, the number in **TCNT1H** is at the same time automatically stored in an internal TEMP register on board the AVR. When you then try to read in **TCNT1H**, the value read is taken from the TEMP register, and not from **TCNT1H**. Note that the internal TEMP register is completely separate to the working register R16 which we often call **temp**.

Example 4.5   Read Timer/Counter 1 into two working registers, **TimeL** and **TimeH**.

Value in T/C1
0x28FF	**in**	**TimeL, TCNT1L**	; **stores FF in TimeL, and stores 0x28**
			; **into the internal TEMP reg.**
0x2900	**in**	**TimeH, TCNT1H**	; **copies TEMP into TimeH**

Therefore, even if T/C1 changes from 0x28FF to 0x2900 in between reading the bytes, the numbers written to **TimeL** and **TimeH** are still 0x28 and 0xFF, and *not* 0x28 and 0x00.

Similarly, when writing a number to both the higher and lower registers *you must write to the higher byte first*. When you try to write a number to **TCNT1H**, the AVR stores the byte in the internal TEMP register and then, when you write the lower byte, the AVR writes both bytes *at the same time*.

Example 4.6   Write 0x28F7 to the Timer/Counter 1.

ldi	**TimeL, 0x28**	;
ldi	**TimeH 0xF7**	;
out	**TCNT1H, TimeH**	; **writes 0x28 into internal TEMP reg.**
out	**TCNT1L, TimeL**	; **writes 0xF7 to TCNT1L and 0x28 into**
		; **TCNT1H at the same time**

The T/C1 has some other 2-byte registers associated with it, such as **ICR1H, L** and **OCR1AH, L**, and they must be written to and read from in the same way as **TCNT1H, L**. The functions of these registers are discussed in the next two sections.

Input capture

Let's say, for example, that we wish to measure the time until an event occurs on a certain pin (as we had to do with the frequency counter project). We could just test the pin and then read the T/C1 as we did before, but in order to simplify the program and free up the processor on the chip, we can use a handy feature that captures the value in T/C1 for us. The *input capture* feature automatically stores the value in T/C1 into two I/O registers: **ICR1H** (Input Capture Register for Timer/Counter 1, Higher byte) and **ICR1L** (Lower byte) when an event occurs on the ICP (Input Capture Pin), which is PD6 on the 2313. This event can be a rising or falling edge. The input capture feature is controlled by an I/O register called **TCCR1B** (one of the two Timer Counter 1 Control Registers) – the other control register for T/C1 is called **TCCR1A** and will be discussed in the next section.

<u>TCCR1B</u> – Timer Counter 1 Control Register B ($2E)

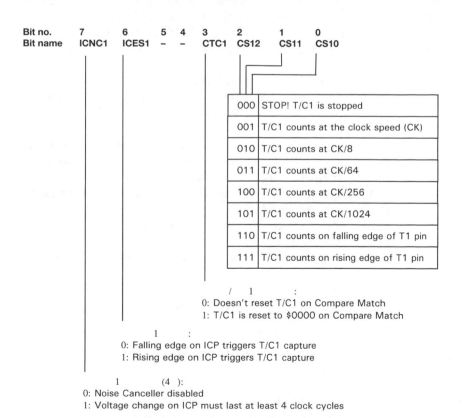

Bit no.	7	6	5	4	3	2	1	0
Bit name	ICNC1	ICES1	–	–	CTC1	CS12	CS11	CS10

000	STOP! T/C1 is stopped
001	T/C1 counts at the clock speed (CK)
010	T/C1 counts at CK/8
011	T/C1 counts at CK/64
100	T/C1 counts at CK/256
101	T/C1 counts at CK/1024
110	T/C1 counts on falling edge of T1 pin
111	T/C1 counts on rising edge of T1 pin

/ 1 :
0: Doesn't reset T/C1 on Compare Match
1: T/C1 is reset to $0000 on Compare Match

1 :
0: Falling edge on ICP triggers T/C1 capture
1: Rising edge on ICP triggers T/C1 capture

1 (4):
0: Noise Canceller disabled
1: Voltage change on ICP must last at least 4 clock cycles

Figure 4.11

Bit 7 can be used to make the feature more robust to noise on the ICP pin. If this feature is enabled, the voltage must rise from logic 0 to 1, for example, and stay at logic 1 for at least four clock cycles. If the voltage drops back to logic 0 before the four clock cycles have passed, the signal is rejected as a glitch, and there is no input capture. If you are trying to read signals that will be less than four clock cycles, you will have to disable this noise cancelling feature (clear the bit). Bit 3 refers to the *output compare* function which is introduced in the next section. There is an input capture interrupt to let us know when an input capture has occurred. This calls address $003 (on the 2313). The enable bit is bit 3 of **TIMSK**.

Example 4.7   Input capturing could be used in a speedometer on a bicycle, where a magnet would pass by the sensor with every revolution of the wheel. The speed of the bike could be deduced as a function of the time between each revolution. The magnetic sensor could be attached to the ICP pin, which would go high every time the magnet passes over the sensor. We would want to be able to measure times up to about 1 second, which means prescaling of the CK/256 would be ideal. You may wish to remind yourself of the 2313 interrupt vector table in Appendix E. The skeleton of a speedometer program is shown below:

```
            rjmp    Init            ; address $000
            reti                    ; $001 – not using INT0 interrupt
            reti                    ; $002 – not using INT1 interrupt

IC_Int:                             ; $003 – the Input Capture interrupt
            in      temp, ICRL      ; stores captured value in working
            in      tempH, ICRH     ;    registers
            sub     temp, PrevL     ; finds different between old and new
            sbc     tempH, PrevH    ;    values
            mov     PrevL, ICRL     ; stores new values
            mov     PrevH, ICRH     ;
            rcall   DigConvert      ; converts two-byte time into digits
            reti                    ;

Display:    etc.                    ; left for you to write
            ret

DigConvert: etc.                    ; left for you to write
            ret

Init:       ldi     temp, 0b11000100    ; enables noise canceller
            out     TCCR1B, temp        ; T/C1 counts at CK/256
            ldi     temp, 0b00001000    ; enables TC interrupt
            out     TIMSK, temp         ;
```

	sei	; **enables global interrupt**
	etc.	; **left for you to write**
Start:	**rcall Display**	; **keeps updating the displays**
	rjmp Start	; **loops**

The display and digit-convert subroutines are not included, but it is expected that you could write them based on the similar display routines in previous example projects. Note that the **DigConvert** subroutine should convert the number held over **temp** and **tempH** (i.e. the difference between the two times) into the digits to be displayed. The remainder of the **Init** section should also be completed – this sets up the inputs and outputs. Note that even though we are not using the interrupts that point to addresses $001 and $002, we still need instructions for those addresses. We could just use **nop** (**n**o **op**eration, i.e. do nothing), but **reti** is safer. The idea is that if by some unforeseeable error an INT0 interrupt is triggered, the program will simply return, and no damage will be done. This is a basic example of *defensive programming* – i.e. expect the unexpected.

Output compare

In almost any application of the timer/counters, you are testing to see if the timer/counter has reached a certain value. Fortunately, all chips with a 'Timer/Counter 1' have a built-in feature which does this automatically. We can ask the AVR to continually compare the value in T/C1 with a certain 16-bit value. When T/C1 is equal to this value, an interrupt can occur, or we can change the state of one of the output pins, and we can also make the T/C1 reset (see bit 3 of the **TCCR1B** register shown on page 119). On the 2313, for example, the value that is to be compared with T/C1 is stored over two I/O registers: **OCR1AH** and **OCR1AL** (which stand for **O**utput **C**ompare **R**egister **A** for T/C1, **H**igher and **L**ower bytes respectively). The 'A' is to distinguish them from a second set of output compare registers (labelled 'B') that are found in other chips such as the 8515. The 8515, for example, can therefore constantly compare T/C1 with *two* different values. If we wish to use the output compare feature we will need to enable the Output Compare Interrupt, which occurs when **TCNT1H** = **OCR1AH** and **TCNT1L** = **OCR1AL**. The enable bit for this interrupt is in **TIMSK**, bit 6. The interrupt address varies between different models, but for the 2313 the output compare interrupt calls address $004. We will find the output compare feature very useful in the next project, and in the next chapter we will see how it can be used for PWM (pulse width modulation).

EXERCISE 4.20   *Challenge!* If we want an interrupt to occur every second, and we are using a 4 MHz oscillator, suggest numbers that should be moved into the following registers: **TCCR1B, TIMSK, OCR1AH, OCR1AL**.

Major program N: melody maker

- EEPROM
- Output compare
- Sounds

By driving a speaker at a certain frequency, we can use the AVR to create musical notes. In fact, using a square wave actually creates a more natural sound than a sine wave input. This end-of-chapter project will allow the user to program short melodies into the EEPROM of the chip, and then play them back through a speaker. The relation between some musical notes and frequencies is shown in Table 4.5.

Table 4.5

C	C#	D	D#	E	F
128 Hz	136 Hz	144 Hz	152 Hz	161 Hz	171 Hz
F#	G	G#	A	A#	B
181 Hz	192 Hz	203 Hz	215 Hz	228 Hz	242 Hz

The values for the next highest octave can be obtained by doubling the frequency. For example, the next 'C' will be at 256 Hz. Assuming we use four octaves, we can encode the note as the letter (which needs 4 bits) and the octave number (which needs 2 bits). The length of the note will be encoded in the remaining 2 bits. Each note in the melody will therefore take up 1 byte of EEPROM. The 2313 has 128 bytes of EEPROM, which means we can store a 128-note melody. If longer tunes are required, a chip with more EEPROM can be used instead, such as the 8515. The note will be encoded as shown in Figure 4.12.

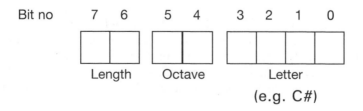

Figure 4.12

The circuit will simply consist of a speaker attached to PB0 (and the usual crystal oscillator on XTAL1 and XTAL2). The AVR can drive a speaker on its own, as long it has a relatively high impedance (e.g. 64 ohm). If you are using a lower impedance speaker (e.g. 8 ohm) you might be better off driving it with a transistor. The flowchart is shown in Figure 4.13; notice how the entire program is interrupt oriented and the main body of the program will simply be a loop.

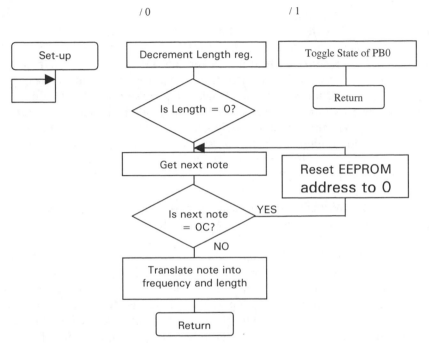

Figure 4.13

A 'note letter' value between 0x0 and 0xB will correspond to a note between 'C' and 'B'. The value 0xC in the 'note letter' part of the EEPROM byte will indicate the end of the melody and cause the chip to return to the start of the melody and repeat it over again. You may want to add extra functionality by including 0xD in the 'note letter' part of the byte, meaning end the melody and *do not* loop back (i.e. just wait until a manual reset), but this is not included in my version of the program. In the Init section, configure the inputs and outputs, the timing registers, and the stack pointer register (**SPL**). Enable the T/C0 Overflow and T/C1 Output Compare interrupts. The T/C1 will be used to create a signal of a certain frequency on the speaker pin, whilst T/C0 will be used to regulate the length of the note. Therefore, set up T/C0 to count at CK/1024, and T/C1 to count at CK. In the Init section you will also have to set up the first note; call a subroutine **Read_EEPROM** to do this, we will write the subroutine later.

At **Start:** you need only write one instruction which loops back to **Start**. Whenever the T/C1 Output Compare interrupt occurs the output will have to change state. This simply involves reading in PortB into **temp**, inverting it, and then outputting it back into PortB.

EXERCISE 4.21   Write the *four* lines which make up the T/C1 Output Compare interrupt section. Include a link to this section at address $004 in the program memory.

All that remains is the T/C0 Overflow interrupt section. **Length** will be a working register we use to keep track of the length of the note. At the start of the section, decrement **Length**. If it isn't zero just return; if it is, skip the return instruction and carry on. If sufficient time has passed, we need to change the note, but first there must be a short pause. This pause allows us to repeat the same note twice without making it sound like a single note played for twice as long. An easy way to insert a pause is simply to wait for the T/C0 Overflow interrupt flag to go high again. If it is, skip out of the loop, reset the flag and move on to the section that reads the next note. Call this section **Read_EEPROM**.

EXERCISE 4.22   Write the *eight* lines at the start of the T/C0 Overflow interrupt section. Include a link to this section at address $006.

The **Read_EEPROM** section copies the number in a working register called **address** into **EEAR**. Read the EEPROM into the **ZL** register, and mask bits 4–7, selecting the 'note letter' part of the byte. Then compare ZL with the number 12 (0xC); if it is equal, jump to a section called **Reset**. If it isn't equal test to see if it is less than 12 (**brlo**). If it isn't less (i.e. it is greater than 12) it is an invalid note letter, and so ZL should be reset to 0x0, for want of a better note. If it is less than 12, skip that instruction.

EXERCISE 4.23   Write the first *eight* lines of the **Read_EEPROM** section.

We will be using ZL to read values from a look-up table in the program memory (using the **lpm** instruction). As you may remember, **lpm** uses the *byte address* of the program memory, rather than the *word address*, so we need to multiply ZL by two (using the **lsl** instruction). The look-up table will start at *word address* 013. We can ensure this using the **.org** directive in AVR Studio. This says 'Let the next instruction be placed at address …'. Our look-up table starts as shown below (**.dw** is the directive which puts the word or words which follow in the program memory).

```
.org  13
.dw  0x7A12        ; frequency for C    (word address 013)
```

.dw 0x7338   ; frequency for C#   (word address 014)
etc.

We must therefore add 26 to ZL to correctly address the look-up table. Use **lpm** to read the lower byte, and move the result from R0 into a working register called **NoteL**. Then increment ZL and do the same, moving the result into **NoteH**.

EXERCISE 4.24   What *seven* lines perform this task?

We will need to perform some basic maths to derive the values for the look-up table. Taking the frequencies of the lowest octave to be played, shown in Table 4.5, and dividing by 4 000 000 (the oscillator frequency) by these values, we get a set of numbers indicating the numbers with which we wish to compare T/C1. To get higher octaves we will simply divide these values by two. My values are shown in the full version of the program in Appendix J; you may wish to check them, or else you can simply copy them.

To get the correct octave we again copy **EEDR** into **temp**, swap the nibbles, and then mask bits 2–7, leaving us with the 2 bits we are interested in – those that choose the octave. Label the next line **GetOctave**. First test if the result of the AND operation just performed is 0; if it is we can just move on to the next section – **GetLength**. If it isn't 0, we will divide the number spread over **NoteH** and **NoteL** by two, decrement **temp**, and then loop back to **GetOctave**.

EXERCISE 4.25   Write the *eight* lines that use bits 4 and 5 of the EEPROM byte to alter the frequency according to a specified octave.

NoteH and **NoteL** are now ready to be moved into **OCR1AH** and **OCR1AL**, but remember to write the *higher byte first*. We then read the length, using a similar method to **GetOctave**. Again read the **EEDR** into **temp**, mask bits 5–0, swap the nibbles, and rotate once to the right. This places the relevant bits in bits 1 and 2 of **temp**. This means the number in **temp** is 0, 2, 4 or 6. This is almost what we want, and by adding 2 to **temp** we get 2, 4, 6 or 8. This should be moved into **Length**.

EXERCISE 4.26   What *nine* lines make up the **GetLength** section and return from the subroutine, enabling interrupts.

The program is now finished. By programming different values into the EEPROM when you program the chip, it can be made to produce any tune. You may find a spreadsheet useful in converting notes, octaves and lengths into the hex number which represents them. You may also want to look into ways to input bytes to the EEPROM more easily. For example, you could use an array of push buttons in a keyboard arrangement, strobing them to lessen the number of inputs needed, to input the melody. Another method might involve a seven segment display to display the note, with a series of buttons to scroll through the memory and change the note – this would require less skill as a pianist to enter the tune!

5
Advanced operations

PWM – pulse width modulation

In this section we will see how the output compare function can be used to create an analogue output – a simplification of the method used in the voltage inverter project. Our aim is to create a square wave output whose *mark–space ratio* we can change. The mark–space ratio is the duration of the 'logic 1' part of the wave divided by the duration of the 'logic 0' part of the wave. By controlling this ratio, we can control the output voltage, which is effectively an average of the square wave output, as shown in Figure 5.1. When using this output, you may need to add a resistor/capacitor arrangement similar to that used in the voltage inverter project, depending on the application.

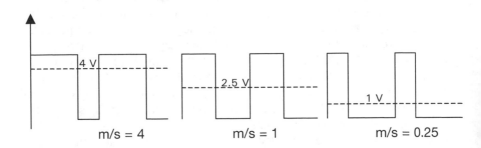

Figure 5.1

The output compare function is used to create automatic PWM, with 8-, 9-, or 10-bit resolution. By placing T/C1 in 8-bit PWM mode, for example, we force T/C1 into a mode whereby it counts up to 0xFF, and then counts back down to 0x00, and then repeats. We then set a threshold by moving a certain number into the output compare registers. When T/C1 reaches this value when counting up, it will set the OC1 output pin (PB3 on the 2313). When T/C1 reaches the value when counting back down it will clear the OC1 output pin. This creates 8-bit PWM, as illustrated in Figure 5.2.

If in 9-bit PWM mode, T/C1 will count up to 0x1FF before counting back down, giving an extra bit of resolution. Similarly, in 10-bit PWM mode, T/C1 will count up to 0x3FF and back. You are also able to invert the PWM output so

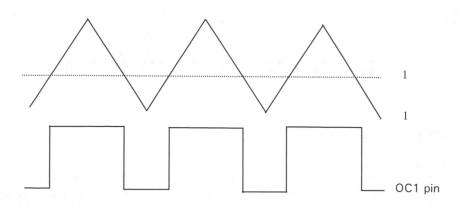

Figure 5.2

that the OC1 is *cleared* when T/C1 passes the threshold whilst counting up, and OC1 is *set* when T/C1 passes the threshold whilst counting down. The I/O register **TCCR1A** controls the PWM settings, the bit assignments are shown in Figure 5.3.

First, you will notice that you have the option, when not in PWM mode, to alter the state of the OC1 pin whenever Output Compare interrupt occurs. We could use this in the melody maker project to toggle the speaker output automatically, if we connected the speaker to OC1. You may also be wondering what happens to the T/C1 Overflow interrupt when in PWM mode (as in this case the T/C1 clearly *never* overflows). When in PWM mode, the T/C1 Overflow interrupt occurs every time T/C1 starts counting from 0x0000. Furthermore, if PWM is enabled, the OC1 is treated as an output, regardless of the state of the corresponding bit in the **DDRx** register.

There is another feature of the PWM mode which comes into effect whenever you try to change the output mark–space ratio. You would do this by changing the **OCR1AH** and **OCR1AL** registers, but unless you change them at precisely the moment at which T/C1 is at its maximum (e.g. 0x1FF for 9-bit PWM), you run the risk of a glitch appearing in your output. This glitch would take the form of a pulse whose width was in between the old and new widths. In cases where you are trying to send information encoded in the length of the pulses, this would clearly be damaging, as you would send some garbage every time you changed the pulse width. Thankfully, in PWM mode, when you try to change **OCR1AH** and **OCR1AL**, their new values are stored in a temporary location, and they are properly updated only when T/C1 is at its maximum.

TCCR1A – Timer/Counter 1 Control Register ($2F)

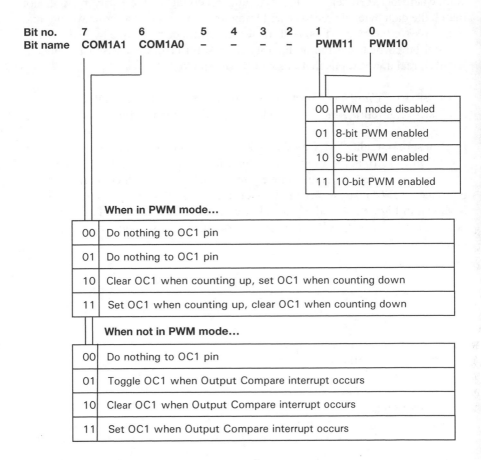

Figure 5.3

UART

'UART' is an Egyptian term that means 'the Artist's Quarter' – a place of bifurcation or division. However, UART also stands for Universal Asynchronous Receiver and Transmitter, and is a standardized method of sharing data with other devices. The UART module found on some AVR models (such as the 2313, 4433 and 8515) refers to the latter. UART involves sending 8- or 9-bit packets of data (normally a byte, or a byte plus a parity bit). This 8- or 9-bit packet is called a *character*. A parity bit is an extra bit sent along with the data byte that helps with the error checking. If there are an odd number of ones in the data byte (e.g. 0b00110100), the parity bit will be 1, if there are an even number (e.g. 0b00110011), the parity bit will be 0. This way, if a bit error occurs somewhere between sending the byte and receiving it, the parity bit will not match the data byte, the receiver will know that something has gone wrong, and it can ask for the byte to be resent. If *two* bit errors occur in one byte, the parity bit will be correct, but the probability of *two* bit errors occurring is often so small in real applications that this can be overlooked.

EXERCISE 5.1   *Challenge!* Write a short piece of code that takes the number in a register (e.g. **temp**), and works out the state of the parity bit for that register.

For transmission, the UART module takes the input character (8 or 9 bits), adds a *start bit* (a zero) at the front, and a *stop bit* (a one) to the end, to create a 10- or 11-bit sequence. This is then moved into a *shift register* which rotates the bits on to the TXD (transmission) pin, for example pin PD1 on the 2313. An example is shown in Figure 5.4, and the speed at which the bits are moved on to the pin is dictated by the *baud rate* (number of bits per second) which can be controlled.

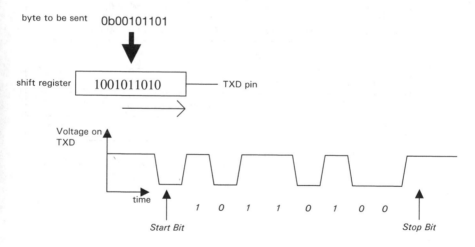

Figure 5.4

The UART module at the receiving end will be constantly checking the data line (connected to the RXD pin), which will normally be high. The receiver can actually sample the data line at *16 times* the baud rate, i.e. it can make 16 samples per bit. If it detects that the RXD pin goes low (i.e. a potential *start bit*) it waits for six samples and then makes three more samples. These should be samples 8, 9 and 10 out of the 16 for any given bit – i.e. it is sampling at the middle of the bit, allowing for slow rise and fall times on the signal. If it detects that the RXD pin is *still* low, i.e. this is *definitely* a start bit, it carries on and reads the whole byte. If the RXD is no longer low, it decides the first sample must have been noise and carries on waiting for a genuine character. If the receiver has decided that this is a genuine character, it will sample each bit three times at the middle of its pattern. If the values of the three samples taken on the same bit are not all identical, the receiver takes the majority value. Finally, when the receiver samples what it thinks should be the *stop bit*, it must read a one (on at least two of the three samples) to declare the character properly read. If it doesn't read a stop bit when it expects to, it declares the character *badly framed* and registers a *framing error*. You should check to see if a framing error has occurred before using the value you have just read into the chip.

Fortunately, all this is done for us by the UART module on the AVR chip. The UART module also brings with it four I/O registers:

UDR (UART Data Register, $0C) – Bits 0 to 7 of the data to be sent, or data just received
UCR (UART Control Register, $0A) – Controls settings of the UART, and contains bit 8
USR (UART Status Register, $0B) – Displays status of parts of UART (e.g. interrupt flags)
UBRR (UART Baud Rate Register, $09) – Sets the speed of the UART data transfer

The bit assignments for registers **UCR** and **USR** are shown in Figures 5.5 and 5.6 respectively.

Finally, **UBRR** is used to control the rate of the data transfer. Clearly, this must be the same for both the transmitting device and the receiving device. This *baud rate* is given by the following formula:

$$\text{Baud rate} \ = \ \frac{\text{CK}}{16 \times (\textbf{UBRR} + 1)}$$

For example, if we are using a 4 MHz clock, and the number in **UBRR** is 25, the baud rate will be about 9615. There are a number of standard values for baud rates: 2400, 4800, 9600 etc., which it can be advisable to stick to, to allow compatibility of your device with others. For this reason, oscillator frequencies such as 4 MHz are not very good for UART applications, as it is impossible to choose these standard values of baud rates (try **UBRR** = 26 in the above). Much better values include 1.8432 MHz, 2.4576 MHz, 3.6864 MHz, 4.608 MHz, 7.3728 MHz, and 9.216 MHz. For the higher frequencies, make sure the AVR model you have chosen can take such a clock frequency. Taking 3.6864 MHz as an example, we can see that **UBRR** = 23 leads to a baud rate of exactly 9600.

Example 5.1 Send the value in the working register **Identity** to another UART device:

UCR – **U**ART **C**ontrol **R**egister (**$0A**)

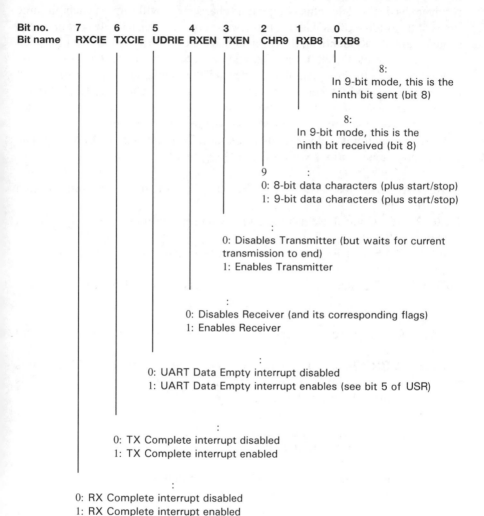

Bit no.	7	6	5	4	3	2	1	0
Bit name	RXCIE	TXCIE	UDRIE	RXEN	TXEN	CHR9	RXB8	TXB8

8:
In 9-bit mode, this is the
ninth bit sent (bit 8)

8:
In 9-bit mode, this is the
ninth bit received (bit 8)

9 :
0: 8-bit data characters (plus start/stop)
1: 9-bit data characters (plus start/stop)

:
0: Disables Transmitter (but waits for current
transmission to end)
1: Enables Transmitter

:
0: Disables Receiver (and its corresponding flags)
1: Enables Receiver

:
0: UART Data Empty interrupt disabled
1: UART Data Empty interrupt enables (see bit 5 of USR)

:
0: TX Complete interrupt disabled
1: TX Complete interrupt enabled

:
0: RX Complete interrupt disabled
1: RX Complete interrupt enabled

Figure 5.5

```
ldi     temp, 0b00001000   ; enables the transmitter
out     UCR, temp          ;
out     UDR, Identity      ; sends value
```

If we wished to send another piece of data, we would have to wait for the **UDRE**

<u>USR</u> – **U**ART **S**tatus **R**egister ($0B)

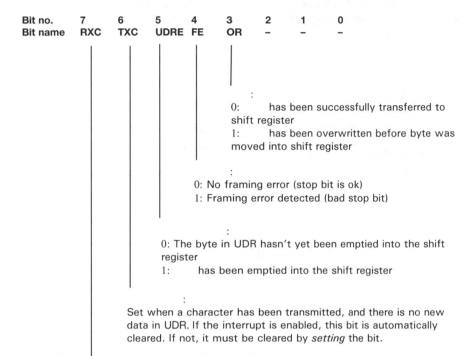

Bit no.	7	6	5	4	3	2	1	0
Bit name	RXC	TXC	UDRE	FE	OR	–	–	–

:
0:      has been successfully transferred to shift register
1:      has been overwritten before byte was moved into shift register

:
0: No framing error (stop bit is ok)
1: Framing error detected (bad stop bit)

:
0: The byte in UDR hasn't yet been emptied into the shift register
1:      has been emptied into the shift register

:
Set when a character has been transmitted, and there is no new data in UDR. If the interrupt is enabled, this bit is automatically cleared. If not, it must be cleared by *setting* the bit.

:
Set when a character has been received and stored in UDR. If the interrupt is enabled,   must be read to clear this bit.

Figure 5.6

bit in **USR** to tell us that the byte has been moved into the shift register, and **UDR** is ready for a new byte.

You can use UART to communicate with the RS232 port on your PC. The simplest way to send bytes through your PC's serial port is through a program that comes with Microsoft® Windows® called HyperTerminal (Start Menu → Programs → Accessories → Communications). You can create a connection with your serial port (e.g. COM1), choose a baud rate, number of bits, parity setting etc. When HyperTerminal connects to the serial port, whatever character you type is sent (as ASCII) through the serial port. If you have a development board, such as the STK500, there is an RS232 socket that you can connect directly to the RXD and TXD pins. If you do not have such a development board, you will have to wire up the correct pins to RXD and TXD, and also make sure the voltage (which could be anywhere between 3 and 12 V), is regulated to a safe voltage (like 5 V). Figure 5.7 shows how to wire up the pins on a 9-pin RS232 socket to allow direct communication with the AVR. Some of the other pins are *handshaking pins*, which can be bypassed by connecting them together as shown.

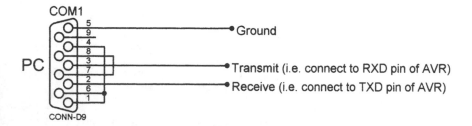

Figure 5.7

Program O: keyboard converter

- UART
- Sounds
- Seven segment displays
- Output compare

We can use HyperTerminal to send characters to our melody maker project, via the UART module. We can effectively convert our computer keyboard into a musical keyboard by assigning note frequencies to the different characters. For example, when I press the letter 'a' when HyperTerminal is connected to the AVR, it will send 'a' to the UART module. This can then trigger an interrupt, convert the ASCII code for 'a' into the frequency for a 'C' note. I have arranged my keys on the keyboard so that they resemble how they are arranged on a piano, but you may find you can fit more notes if you arrange them differently. Figure 5.8 shows my arrangement.

Figure 5.8

We will also use a seven segment display to show which note is being played; this can help overcome any user confusion over how the letters on the computer keyboard correspond to the musical notes. There will be a separate LED to show the sharp symbol (#). The circuit diagram is shown in Figure 5.9, and the flowchart in Figure 5.10.

In the Init section, set up inputs and outputs and set OC1 to toggle with every output compare (this handles the speaker output for us, so there is no need to write a routine for the Output Compare interrupt). Make all other timer settings the same as in the melody maker, choose a baud rate of 9600, and enable the UART receiver and the UART Receive Complete interrupt.

Again, the main body of the program is just a constant loop to **Start**. The UART Receive Complete interrupt tells us that some new data has been received on the line, which we should convert to a frequency and then change **OCR1AH** and **OCR1AL** accordingly. The beginning of the interrupt routine should therefore read **UDR** into ZL. The ASCII conversion table is shown in Appendix G. I will only use letters a–z, all lower case, which correspond to 0x61 to 0x7A in ASCII, so subtract 0x61 from ZL to get a number between 0 and 25. If ZL is more than 25, an inappropriate key is being pressed, so move 26 into it, this ensures no matter what character we read, the program will stay within the look-up table we are about to write. Now multiply ZL by two to make it a word address. We wish to read the program memory into R0, using the **lpm** instruction, and then copy R0 into **OCR1AH**, and **OCR1AL**. We can do this directly (i.e. without having to play with octaves etc., so we don't need **NoteH** and **NoteL**). However, when doing this directly, we have to remember the golden rule – you must write the higher byte first. There are two ways of doing this. First, arrange the data in the look-up table so that the higher byte actually comes first. For example, if I wished the number 0x1E84 to be the code for a 'C' note, the top of my look-up table would be:

.dw   0x841E

This is a little confusing, and an easier way is to start by pointing ZL to the higher byte. In other words, if the table starts at byte address 26 in the program memory, add 27 to ZL instead of 26, to point ZL to the higher bytes. Then to read the lower byte, *decrement* ZL.

EXERCISE 5.2   *Challenge!* Write the first *12* lines of the UART Receive Complete interrupt section which use the data received by the UART module to write new values for **OCR1AH** and **OCR1AL**.

For the display we have another look-up table, below the first, starting at word address 43. We can simply add 60 (30 × 2) to ZL to point to the second look-up table. This holds the seven segment codes for the note letters. Bit 3 will be used to light up the # (sharp) LED.

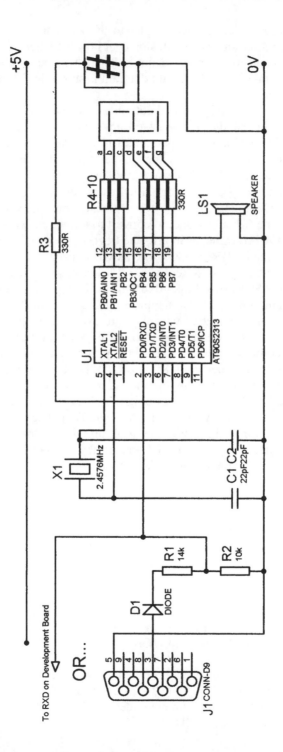

Figure 5.9

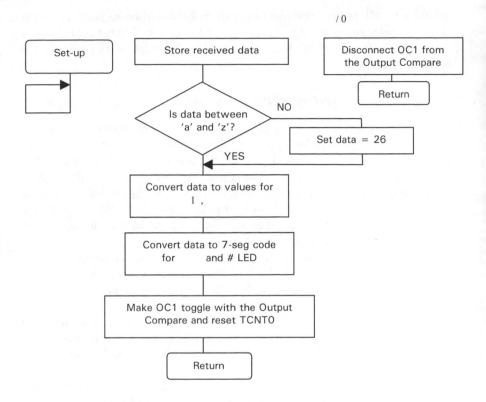

Figure 5.10

EXERCISE 5.3   What *six* lines point ZL to the second look-up table, read the value, and output it to **PortB**? They should then mask all of R0 (which contains the value read from the table) except bit 3, and move the result into **PortD**, to take care of the # LED. As you cannot use the **andi** instruction on R0–R15, you will have to copy R0 into **temp**.

EXERCISE 5.4   What *five* lines will set the OC1 pin to toggle with every Output Compare interrupt, reset T/C0 and return?

EXERCISE 5.5   What *three* lines make up the T/C0 Overflow interrupt, which should disconnect the OC1 pin from the Output Compare interrupt and return?

This program is quite fun to play around with, but you may find the keyboard's *repeat delay* a nuisance. You can try to minimize this in the Control Panel, or perhaps lengthen the minimum note to try to overcome it. If you move the frequencies produced out of the audible range, this project can be developed into more sinister applications – perhaps you could use it for espionage purposes …?

Another UART project you may wish to make would be to build upon the palindrome detector designed in Chapter 3, and interface it with a computer via its serial port. The use of the Receive Complete interrupt would simplify the program considerably.

Serial peripheral interface (SPI)

The UART described in the previous section has a few drawbacks. For a start it is only *half duplex* (also called *simplex*) – this means you can send data in only one direction on one line. Connecting the TXD pin on one device connected to the RXD pin of another supports data transfer in one direction only, namely TXD to RXD. SPI offers *full duplex* – the ability to send data in both directions at the same time. It is also a *synchronous* mode of transfer – this means all the relevant devices are also connected to a common clock, so that they can all be in synch, and operate at a higher speed.

Sending information through the SPI module is just as straightforward as with UART. Any number of SPI devices can be connected together; however, one device is called the Master, and the other devices are Slaves. The Master can talk to the Slaves, and the Slaves can talk to the Master, but the Slaves cannot talk to each other. The Master provides the clock that synchronizes the connection, and it decides when it is going to talk to the Slave, and when the Slave can talk to it. Figure 5.11 shows an arrangement with one Master and two Slaves.

When you move a number into the SPI data register of the Master device, it will immediately start a clock signal on the SCK pin (**SPI Clock**), and begin shifting the data out on the MOSI pin (**Master Out, Slave In**) to the Slaves on

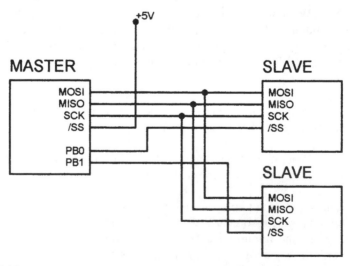

Figure 5.11

their MOSI pins. The Slave will receive the data only if it has been chosen by the Master, i.e. if its $\overline{SS}$ pin is high. Therefore, using any two output pins (PB0 and PB1 in the example in Figure 5.11), the Master can choose which of the Slaves it wants to talk to. As the Master sends its data to the Slave on the MOSI pin, the Slave immediately begins sending the contents of its data register to the Master on their MISO (**M**aster **I**n, **S**lave **O**ut) pins. The two 8-bit shift registers on Master and Slave behave like one big, circular 16-bit shift register – as bits shift off Master onto Slave, bits shift off the Slave and into the Master. You can configure the $\overline{SS}$ pin on the Master as an output, and use it as a general output. If you make it an input, however, you must tie it to V_{CC}, as shown. If the Master's $\overline{SS}$ pin is pulled low, it assumes some other Master wants to enslave it, and will turn into a Slave! This allows some hierarchy between Masters in a complex SPI system. The I/O registers involved with SPI are:

SPDR (**SP**I **D**ata **R**egister, $0F) – Data to be sent, or data just received
SPCR (**SP**I **C**ontrol **R**egister, $0D) – Controls settings of the SPI
SPSR (**SP**I **S**tatus **R**egister, $0E) – Displays status of parts of SPI (e.g. interrupt flags)

SPDR is the data register into which you should move the byte to be sent to the other device, and holds the received byte after the transmission is finished. You must wait for the current transmission to finish before writing the next byte to be sent to **SPDR**. When reading the received byte, you have slightly longer to read it. You can read the received byte while the next transmission is in progress, but once this next byte is completely received, the old received byte is over-written. You therefore have until the next transaction *completes* to read the received data.

The **SPSR** contains two flags. Bit 6 is the *write collision flag*, which is set when **SPDR** is written to before the current transmission is finished. Bit 7 is the *SPI interrupt flag*, which is set when an SPI transmission completes.

An example project you may wish to consider attempting could be an electronic chess game involving two AVR units which communicate using an SPI link. The users at either end can input their move into their unit, which will then send the move to the other unit. The game can be stored on the EEPROM (thus allowing games to continue after power has been removed and the units separated). Sixty-four bytes are required, as each square on the board can be assigned a space in the EEPROM. The number in the EEPROM indicates which piece is on that space. For example 00 could mean empty. 01 = black pawn, 02 = black knight etc., 81 = white pawn, 82 = white knight etc. The allowed moves would involve adding or subtracting numbers to a particular piece's position. For example, allowed moves for bishops are at the basic level adding or subtracting multiples of 9 or 7. Figure 5.13 should help you picture this. However, tests will be needed to ensure the piece doesn't travel through another, or off the board.

SPCR – SPI Control Register ($0D)

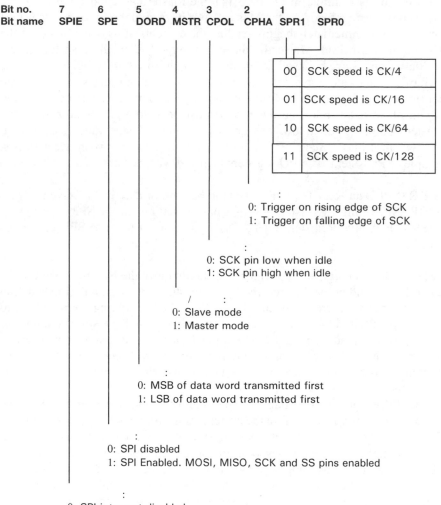

Bit no.	7	6	5	4	3	2	1	0
Bit name	SPIE	SPE	DORD	MSTR	CPOL	CPHA	SPR1	SPR0

00	SCK speed is CK/4
01	SCK speed is CK/16
10	SCK speed is CK/64
11	SCK speed is CK/128

0: Trigger on rising edge of SCK
1: Trigger on falling edge of SCK

0: SCK pin low when idle
1: SCK pin high when idle

0: Slave mode
1: Master mode

0: MSB of data word transmitted first
1: LSB of data word transmitted first

0: SPI disabled
1: SPI Enabled. MOSI, MISO, SCK and SS pins enabled

0: SPI interrupt disabled
1: SPI interrupt enabled

Figure 5.12

The moves could be entered in standard chess notation (e.g. Be2 = Bishop to the E2 square), or with the help of a more visual display which resembles the board. This project is left as an exercise for the chess enthusiasts, but I would be interested in seeing your solutions (my email address is given in Appendix I).

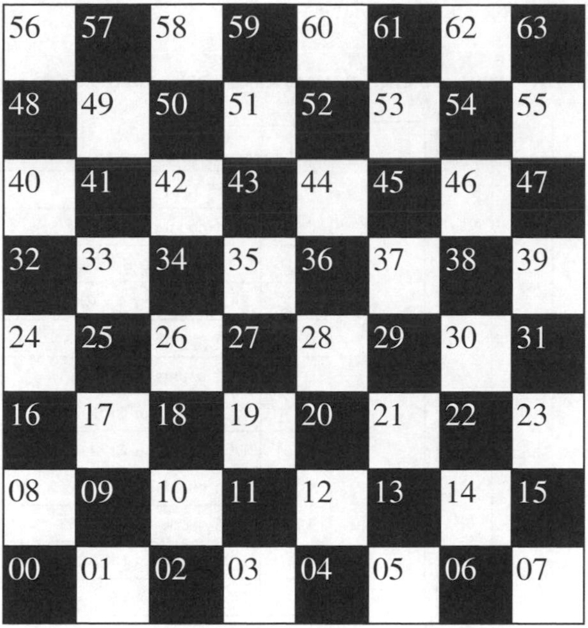

56	57	58	59	60	61	62	63
48	49	50	51	52	53	54	55
40	41	42	43	44	45	46	47
32	33	34	35	36	37	38	39
24	25	26	27	28	29	30	31
16	17	18	19	20	21	22	23
08	09	10	11	12	13	14	15
00	01	02	03	04	05	06	07

Figure 5.13

Both UART and SPI can be implemented on chips without these custom modules, entirely with software. For more information on these, you can check out Claus Kühnel's book listed in Appendix I, but my advice would be simply to use a chip that has the hardware you require.

Tiny15's eccentric timer 1

As a brief aside, it is worth noting that the Tiny15 has an 8-bit T/C1, and a few other eccentricities that make it different from the norm. Whereas on other chips, T/C0 and T/C1 can count up at no more than CK, the clock speed at which instructions are performed, the T/C1 on the Tiny15 can actually count up *faster* than CK. It can be set to count at 16CK, 8CK, 4CK or 2CK, as well as CK, and also at a larger range of fractions of CK, as shown in the Tiny15's bit assignment of **TCCR1**, the **T/C1 C**ontrol **R**egister (Figure 5.14). The reason it can count higher than CK is that it has access to a high-speed clock (called PCK) that runs 16 times faster than CK; values such as 8CK and 4CK are obtained by prescaling this high-speed clock.

As T/C1 is only 8 bit, the PWM is 8 bit. Rather than counting up and down in PWM mode, T/C1 is always counting up, and will change the state of the OC1 pin when it reaches the top. The top value of T/C1 is given by the **OCR1B**

TCCRI – **T/CI C**ontrol **R**egister (**$30**) on the Tiny15

Bit no.	7	6	5	4	3	2	1	0
Bit name	CTC1	PWM1	COM1A1	COM1A0	ADIE	ADPS2	ADPS1	ADPS0

0000	STOP! T/C1 is stopped
0001	T/C1 counts at 16 x CK
0010	T/C1 counts at 8 x CK
0011	T/C1 counts at 4 x CK
0100	TC/1 counts at 12 x CK
0101	T/C1 counts at CK
0110	T/C1 counts at CK/2
0111	T/C1 counts at CK/4
1000	T/C1 counts at CK/8
1001	T/C1 counts at CK/16
1010	T/C1 counts at CK/32
1011	T/C1 counts at CK/64
1100	T/C1 counts at CK/128
1101	T/C1 counts at CK/256
1110	T/C1 counts at CK/512
1111	T/C1 counts at CK/1024

...   ...

00	Do nothing to OC1 pin
01	Do nothing to OC1 pin
10	Clear OC1 when compare match, set on T/C1 overflow
11	Set OC1 when compare match, clear on T/C1 overflow

...   ...

00	Do nothing to OC1 pin
01	Toggle OC1 when Output Compare interrupt occurs
10	Clear OC1 when Output Compare interrupt occurs
11	Set OC1 when Output Compare interrupt occurs

:
0: PWM disabled
1: PWM enabled (8-bit)

/   1   :
0: Doesn't reset T/C1 on Compare Match
1: T/C1 is reset to $00 on Compare Match

Figure 5.14

I/O register. The PWM is glitch free, as before, so updates to **OCR1A** occur only when T/C1 reaches the top value, as shown in Figure 5.15.

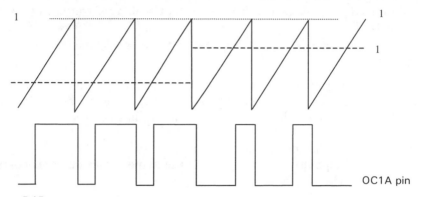

Figure 5.15

As if this wasn't enough, there's another I/O register thrown in, with the mysterious title of **S**pecial **F**unction **IO R**egister: **SFIOR** ($2C). This register allows you to *reset the prescaler* of either of timer/counters. What on earth does this mean? Let's look at how the prescaler works. Essentially, the prescaler is a 10-bit register that counts up at CK. When T/C0, for example, is 'prescaled at CK/2' it counts with bit 0 of the prescaler. If it is 'prescaled at CK/64', it counts with bit 5 of the prescaler etc. This is illustrated in Figure 5.16.

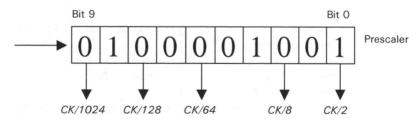

Figure 5.16

When you reset the prescaler, you wipe its value to 0, ensuring a more accurate count. Say you wished to set your T/C0 to count at CK/1024. In *steady state* operation it will be perfectly accurate, but for that very first count, we don't know that the number in the prescaler doesn't happen to be 1023, and so the first count will come a lot sooner than expected. To reset the prescaler for T/C0, just set bit 0 of **SFIOR** (the bit will then clear itself). To reset the prescaler for T/C1, set bit 1 of **SFIOR**. Finally, with bit 2 of **SFIOR**, we are able to force a change on the OC1A pin, according to the settings in bits 4 and 5 of **TCCR1**. In other words, we 'fool' the pin into thinking there has been an Output

Compare Match; however there is *no* interrupt generated, and T/C1 will *not* reset.

Although at the time of publication, the Tiny15 was the only model with this type of T/C1, we can expect that other models of AVR will emerge with a similar T/C1.

Shrtcts

There are a number of ways to trim down your program into a slender and seductive beauty. One of the easiest ways is to use the **.macro** assembler directive. This allows you to, in effect, create your own instructions.

Example 5.1   At the top of your program …
```
.macro      nopnop                    ; the name of this macro is nopnop
            rjmp    PC+1
.endmacro
```

Then, in the rest of your program, you can write the instruction **nopnop**, and the assembler will interpret this as **rjmp   PC+1**. Why have I called this **nopnop**? Jumping to the next line with the **rjmp** instruction wastes *two* clock cycles, as the **rjmp** instruction takes twice as long as most instructions. Writing **rjmp   PC+1** is therefore equivalent to writing two **nop**s, but only takes up one instruction. Macros can also be given operands, which are referred to as @0, @1 etc.

Example 5.2

```
.macro      multiply                  ; the name of this macro is multiply
            mov     temp, @0          ;
            clr     @0                , wipes answer register
            tst     @1                ; tests multplier
            breq    PC+4              ;
            add     @0, temp          ; adds multiplicand to itself
            dec     @1                ;
            rjmp    PC-4              ; repeats
.endmacro
```

In the program, if we wanted to multiply the number in **Seconds** by the number in **Counter**, we could simply write:

multiply   Seconds, Counter

Note that we can use labels in the macro, these will immediately be translated as relative jumps, and so there will be no risk of label duplication should the macro be used more than once in the program.

EXERCISE 5.6   Create a macro called **skeq** which skips the next instruction if the zero flag is set.

EXERCISE 5.7   Create a macro called **HiWait** which will wait until a bit in an I/O register goes high.

It is important to clarify in your mind the distinction between subroutines and macros. Macros are simply ways of abbreviating longer or less pretty pieces of code into neat one-word actions. The assembler will expand these out, so your program will end up just as long (but you will never see the expanded version). Using subroutines will actually make your program shorter (i.e. take up less space in the program memory), BUT may well take longer to run. The **rcall** instruction takes three clock cycles, and the **ret** instruction four clock cycles, so subroutines are literally a waste of time for really short shortcuts.

A Mega summary

Covering the cornucopia of new functions found on the MegaAVR range is not one of the aims of this book. It is worth, however, giving a brief introduction so that you can at least decide whether it's worth learning more about them. First, they offer more of what you've seen so far: more timers, more PWM, more ADCs, more I/O pins, more memory and more instructions.

The new instructions fall into three categories. There are a few new instructions introduced along with an *on-chip multiplier* – specially built hardware which performs multiplication in two clock cycles. The **mul** instruction is used to multiply two registers together. Other multiply instructions (signed/unsigned/fractional etc.) are also available. The **call** and **jmp** instructions are direct calls and jumps respectively. The only difference to the user is the ability to jump to, or call, *any* part of the program, though you probably won't experience this limitation on non-Mega AVRs unless you write really large programs. The new instructions also include additions to the memory access instructions, most notable is the **stm** instruction. This **st**ores the *word* spread over R0 and R1 into the program **m**emory. This allows the program to write to itself!

Another particularly useful feature available on most new AVRs is the *JTAG interface*. This is a standard that has been developed to facilitate debugging. It is a way for the AVR to send the entire contents of its registers (I/O, working registers, SRAM) to a PC, so that you can see what's going on inside it as it runs in your circuit board.

Final program P: computer controlled robot

- Serial communication
- PWM to drive a motor
- Seven segment display to display messages

A computer controlled robot has been chosen as a fun project which ties together some of the topics discussed in the book. The project that will be developed will be a skeleton, around which a semi-intelligent robot can be based. We can send commands to the robot through the serial port on the computer to the UART module on the AVR. Motor speed can be controlled through the use of PWM, and a seven segment display will be used to show messages, and allow the robot to 'talk'. The use of EEPROM to store moves and the application of the music modules are some basic enhancements that could be added on. Sensors could be placed on the robot, and it could send information back to the computer regarding the states of these sensors. More sophisticated software on the computer end, which would make the robot behave like a state machine and respond to various inputs, would be a more interesting development, but this goes beyond the scope of this book. The circuit diagram of the basic robot is shown in Figure 5.17.

Both motors are driven from the OC1 pin, which is the output of the PWM. To allow the robot to turn, the left motor can be turned off by setting the PD2 pin. This means it can turn in one direction only, but still gives it plenty of freedom. A larger AVR, such as the 8515, has two PWM outputs, on OC1A and OC1B pins. This means the motors can be driven independently.

The commands we can send the robot are shown in Table 5.1.

Table 5.1

Letter	ASCII	Function	Message to PC
g	0x67	Go/Stop	'Go' or 'Stop'
t	0x74	Begin turning or end turning (stop/start left motor)	'Turning'
+	0x2B	Speed up	'Speeding up'
-	0x2D	Slow down	'Slowing down'
s	0x73	Change speed (followed by two-digit number, e.g. **s25**)	'Speed set to ...'
[	0x5B	Begin message (to be displayed on seven segment displays)	<message>
]	0x5D	End message	

All other inputs will be ignored. The robot will send the computer back confirmations of each action. For example, if it is sent a '**t**', it will reply with 'Turning'. Not all letters can be displayed on the seven segment displays – to be able to display any letter we need a more complex display (e.g. a 14 segment display). As it is, we are unable to display letters k, m, q, v, w and x.

The structure of the program is very straightforward, and entirely interrupt driven. If a receive interrupt occurs, the program identifies the character received and responds accordingly. To simplify the **Display** subroutine, we can make this

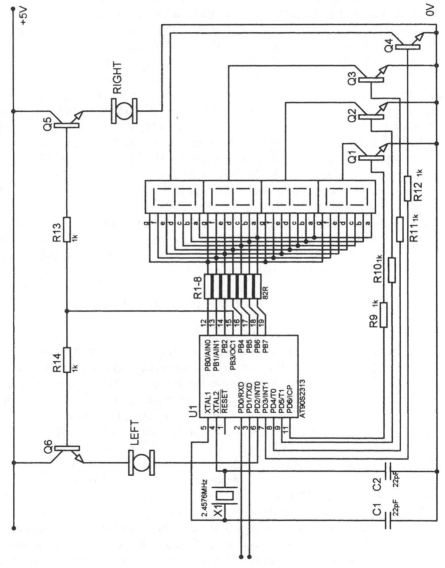

Figure 5.17

driven by T/C0, such that every time T/C0 overflows, the **Display** subroutine is called. This not only removes the burden on us of remembering to call it regularly, but also means we can remove the counter register that allows the entire subroutine to be executed only once every 50 visits. We must therefore configure T/C0 so that it overflows sufficiently often. The refresh rate should be more than 25 times a second and, bearing in mind there are four displays, this means the **Display** subroutine should be called at least 100 times a second. As T/C0 overflows after 256 counts, this means a minimum T/C0 rate of 25.6 kHz. If we are using a 2.4576 MHz crystal, this represents prescaling of CK/64.

In the Init section, configure the inputs and outputs, and T/C0. Set up T/C1 to count at CK, set OC1 to clear when T/C1 passes the threshold counting up, and set when T/C1 passes it coming down (this means the higher the number in **OC1AH/L**, the faster the speed of the motor). Disable PWM for the time being (8-bit PWM will be enabled when a 'g' is received from the computer). Don't forget to set up the stack pointer I/O registers. On the 2313 this is just **SPL**, and which you should load with 'RAMEND'. Enable the Receive Complete UART Interrupt, and enable the Receive Mode. Set the UART baud rate to 9600, and enable the global interrupt bit.

Adjust the **Display** subroutine from previous projects to include *four* displays. The seven segment code to be displayed will be stored in registers **R21–24**. Note that as these will hold the seven segment code, their values can be moved directly into PortB.

EXERCISE 5.8   Make the necessary changes to create a **Display** subroutine for this program.

The Receive Complete Interrupt should first test to see if what is being sent is to be taken as a command, or as part of a text message. The T bit will be used to indicate which interpretation is appropriate (i.e. the start message command '[' will set the T bit, and the end message command ']' will clear it. It should also be cleared in the Init section. The Receive Complete Interrupt section should start by testing for an end message symbol, and jump to **EndMessage** if it is received. The next test should be the T bit, if it is set we should branch to **Message**. The other symbols (g, t, s, +, –) can be tested in any order, though it is simplest to put the test for '[' at the end. If it is '[', the T bit should be set. Any other symbol should be ignored.

The **Turning** section should toggle the state of the PD2 pin (which controls the left motor). The receive mode should then be disabled, and the transmit mode enabled. Move the ASCII code for a 'T' into **temp**, and then call a subroutine called **Send**. This subroutine will take the number in **temp** and send it through the UART module; we will write the subroutine later. Repeat the above for the rest of the letters. We also need to send a *new line* (also called *line feed*) and *carriage return* symbol, so that each message sent to the PC appears on a new line. These symbols are 0x0A and 0x0D respectively, but these will be common to all messages, so at this point (after sending the 'g'), just branch to **EndMessage**, which will do the rest.

EndMessage will clear the T bit, send 0x0A and 0x0D to the PC, and then disable the transmit mode and enable the receive mode.

The **Send** subroutine should put the contents of **temp** into the **UDR**, and then enter a loop in which it constantly checks the transmit complete flag (the TXC bit in **USR**). You must not write to UDR in this loop (i.e. loop to **Send+1**, and not to **Send)**, because this resets the TXC flag, which means you will stay in the loop forever. After the TXC flag goes high, you must reset it by *setting* it, and then return.

The **SpeedUp** section will read in the number currently in OCR1AL, and add 10 to it. If the carry flag is set, the number should be capped at 0xFF, and then moved back to OCR1AL. Note that you *cannot* use the following:

> **subi   temp, -10**

This really adds 246 to **temp**, which will almost invariably set the carry flag. You should therefore move 10 into another working register, and add it to **temp** using the **add** instruction. Alternatively, you could use ZL, and the **adiw** instruction. You should then repeat the same steps as in **Turning** to send the appropriate message back to the PC. Similarly, the **SlowDown** section subtracts 10 from OCR1AL, forcing the value to 0 if it goes negative. The usual method is used to send the reply to the PC.

The **GoStop** section is slightly harder. You must first test the state of the PWM (i.e. is it enabled?) by testing bit 0 of **TCCR1A**. If it is enabled, disable it, and send 'STOP!' to the PC. If it is enabled, jump to a different section called **Go**. This section should enable 8-bit PWM (set bit 0 of **TCCR1A**), and send 'GO!' to the PC.

The **ChangeSpeed** section has to wait for two more characters (the two digits of the speed). It should start with a loop to wait for the first character (waiting for the RXC bit in **USR** to set). The first digit received should be moved from the **UDR** into a working register called **speed10**. This number should be copied into a temporary register, and have 0x30 subtracted from it. This converts the ASCII for 0–9, into the numbers 0 to 9. The result of this should then be multiplied by 10, as this is the tens digit. The next digit should then be received, and the result stored in a register called **speed1**. Again, convert this into the actual number (subtract 0x30), and add it to the tens digit. It is important you keep **speed10** and **speed1** unchanged, as these will be used when replying to the PC. The value representing the total two-digit number will be between 0 and 99. We would like to convert this to something between 0 and 255 – an easy way to do this is to multiply it by 3, but cap anything that goes above 255. The result should be moved into **OCR1AL**. The reply should be sent to the PC 'Speed Set To xx', with xx being the new two-digit speed. For letters, we move the ASCII values into **temp** as before. For the actual speed, just copy **speed10** or **speed1** into **temp**, and call **Send**, as before. After sending **speed1**, this section should jump to **EndMessage**.

Finally, the hardest section is **Message**. This converts input characters from ASCII into seven segment code, and scrolls the result through the displays as they come in. The display registers will be called **Thousands, Hundreds, Tens** and **Ones**. As new numbers come in, **Hundreds** will be copied to **Thousands, Tens** to **Hundreds, Ones** to **Tens**, and finally the new number will be written to **Ones**. First, however, we must convert ASCII to seven segment numbers. We will try to display the digits '0' to '9' only, the lower case letters 'a' to 'z', and the upper case letters 'A' to 'Z', with the exclusions we noted earlier. With the letters, where a lower case letter is not possible whilst an upper case is (e.g. 'e' and 'E'), the upper case alternative is returned. This ensures that the program will try to produce the intended case, but gives getting the letter right at all a higher priority. As you may have guessed, this conversion process is carried out with one large look-up table. The first task is simply to reply to the PC with the character just received. This is straightforward – read **UDR** into **ZL**, disable received mode, enable transmit mode, copy **ZL** into **temp**, and then call the **Send** subroutine. Change back into receive mode and disable transmit mode, and then subtract 0x10 from **ZL**. The digits 0–9 start at 0x30 in ASCII, so subtracting 0x10 will make a '0' correspond to 0x20 etc. This is a byte address, so the word address will be half of this, i.e. a '0' corresponds to word address 0x10. We can make this the start of our look-up table (use **.org 0x10** at the start of the table). The first five words in the look-up table can represent the digits 0–9. Make sure you work out your own values for the look-up table, instead of copying those in my program, as your circuit board may not be the same as mine. Capital letters 'A' to 'Z' start at ASCII value 0x41. Rather than writing empty lines into the look-up table, simply write **.org 0x18**, to point the next part of the look-up table at program address 0x18, which is byte address 0x30, which corresponds to ASCII 0x40. The first byte in the table is therefore not important, but the second should correspond to 'A', and so on. Finally, letters 'a' to 'z' begin at ASCII value 0x61, and so use **.org 0x28** at the top of the look-up table for the lower case letters.

I realized when testing that a space (i.e. pressing the space bar) was an important symbol to transmit. This is 0x20 in ASCII, which gets reduced to byte address 0x10, and word address 0x08. A clever way to deal with spaces, therefore, is to make address 0x08 a **nop** instruction (**nop** is translated as 0x0000 by the assembler). **nop** would be read as any of the other bytes, and return 0b00000000 which corresponds to all bits off (i.e. a space). 0x08 happens to be the UART Empty interrupt, which we are not using, so it is fine to simply write **nop**. In the unforeseeable event that the UART Empty interrupt *does* occur, all that will happen is that it will execute the **nop**, and then the **reti** instruction which follows at address 0x09. The program is therefore still immune to an unexpected occurrence of the UART Empty interrupt. Once the program memory has been read, and the values in the registers shifted along, the **Message** section is finished.

This concludes the final program, my version is shown in Program P. I hope you do try to build this one, and work on some enhancements to make it more robot-like. It really is a good platform for a variety of interesting projects.

Conclusions

When you are debugging your own programs, I suggest the following. First, try to break down your program into discrete units which can be tested independently – this way you can pinpoint bugs quickly. Another frustrating problem can be not being able to look inside the register of the AVR while it is running. This can be overcome by using an emulator, though there is a cheaper way. At certain points in the program you could try sending the contents of certain registers through the UART to your PC, and see how they are changing. The insertion of a UART transmission module in your program may not be worth the extra work, but it does give you a good indication of what's going on *inside* your AVR – like a poor man's JTAG or emulator.

Throughout this book we have encountered examples of attempting to perform a task with limited means, and then learning about new tools which allow us to perform these tasks with greater ease. It is often the case that the more complicated the microcontroller becomes, the simpler a given program will become. This gives us some insight into the compromise that chip designers face between giving a chip functionality and keeping it relatively simple. This simplicity is necessary not only to keep costs low, but also to make the chip easy to get to grips with. I have no doubt that new features will emerge on new models of AVR that appear after the publication of this book. These will almost inevitably centre around some I/O register, perhaps with a certain bit assignment that controls different aspects. This information can be gleaned from the chip's datasheets, which should not be as daunting now as they might have been when you started. By reading through these you should be able to keep abreast of any new functions – make sure you keep up to date with these, they're there to make your life as a programmer easier!

Appendix A
Specifications of some PICs

Device	Pins	I/O	ROM	RAM (bytes)	EEPROM (bytes)	Features
Tiny11	8	6	1K	-	-	8-bit timer, WDT, Analogue comparator, 4 interrupts, on-chip oscillator
Tiny12	8	6	1K	-	64	As Tiny11, 5 interrupts
Tiny15	8	6	1K	-	64	As Tiny11, Two 8-bit timers, 4 ADC channels, 8 interrupts, PWM
1200	20	15	1K	-	64	As Tiny11, 3 interrupts
2313	20	15	2K	128	128	Extended instruction set, 10 interrupts, UART, 8-bit and 16-bit timers, PWM, WDT, Analogue comparator
2323	8	3	2K	128	128	Extended instruction set, 2 interrupts, 8 bit timer, WDT
2343	8	4	2K	128	128	As 2323, on-chip oscillator
4433	28	20	4K	128	256	Extended instruction set, 14 interrupts, SPI, UART, 8-bit and 16-bit timers, PWM, WDT, Analogue comparator, six 10-bit A/D channels
8515	40	32	8K	512	512	Extended instruction set, 11 interrupts, SPI, UART, 8-bit and 16-bit timers, 2 PWM, WDT, Analogue comparator
8535	40	32	8K	512	512	As 8515, 15 interrupts, two 8-bit timers, 3 PWM, RTC Timer, eight 10-bit A/D channels

Appendix B
Pin layouts of various AVRs

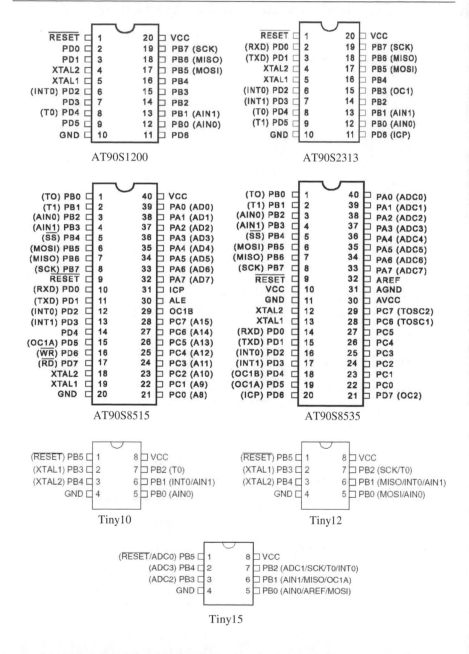

AT90S1200

AT90S2313

AT90S8515

AT90S8535

Tiny10

Tiny12

Tiny15

Appendix C
Instruction overview

BRANCHING

SUBROUTINES

PUSH
pushes reg. onto stack

POP
pops reg. off stack

CALL   long call

ICALL   indirect call

RCALL   relative call

RET   return

RETI   return, enabling interrupts

JMP   long jump

IJMP   indirect jump

RJMP   relative jump

BRBC   branch if SREG bit is clear

BRBS   branch if SREG bit is set

I/O Registers

CBI   †clears IFR bit

SBI   †sets IFR bit

IN   moves IFR into reg.

OUT   moves reg. into IFR

SBIC   † skip if IFR bit is clear

SBIS   † skip if IFR bit is set

SBRC   skip if register bit is clear

SBRS   skip if register bit is set

SREG

BCLR   clear SREG bit

BSET   set SREG bit

BLD   load bit from T

MISCELLANEOUS

NOP   no operation - waste a cycle

WDR   reset watchdog timer

SLEEP   sends chip to sleep

RAM

LD   indirect load from SRAM

ST   indirect store to SRAM

LDS   direct load from SRAM

STS   direct store to SRAM

LPM   indirect load from Prog. Mem.

ARITHMETIC

ADC	adds two regs with carry
ADD	adds two registers
ADIW	adds immediate to word
DEC	decrements register
INC	increments register
LDI	*loads immediate to register
MUL	multiplies two registers
SBC	subs two regs with carry
SBCI	*subs immediate w/ carry
SBIW	subs immediate from word
SUB	subtracts two registers
SUBI	*subs immediate from reg.
COM	inverts all bits of register
NEG	changes sign of register
CLR	clears register (makes 0)
SER	*sets all bits in register
SWAP	swaps upper and lower nibbles

LOGIC

AND	ANDs two regs
ANDI	*ANDs immediate with reg.
EOR	EORs two registers
OR	ORs two registers
ORI	*ORs immediate with reg.

SHIFTING BITS

ASR	arithmetic shift right
LSR	logical shift right
LSL	logical shift left
ROL	rotate left thru carry
ROR	rotate right thru carry

COMPARING

CP	compare two registers
CPC	compare regs and carry
CPI	*compare with immediate
CPSE	compare and skip if equal

Instructions in grey are not available on all chips
* These instructions only operate on working registers R16-R31
† These instruction only operate on I/O registers $00-$1F

Non-critical instructions

Instruction	Action	Equivalent Instruction
CBR Rd, 0bxxxxxxxx	Clears certain bits in a register	ANDI Rd, 0bxxxxxxxx
SBR Rd, 0bxxxxxxxx	Sets certain bits in a register	ORI Rd, 0bxxxxxxxx
TST Rd	Test for zero or minus	AND Rd, Rd
BRCC <label>	Branch to <label> if C flag is clear	BRBC 0, <label>
BRCS <label>	Branch to <label> if C flag is set	BRBS 0, <label>
BRSH <label>	Branch if same or higher (C flag clear)	BRBC 0, <label>
BRLO <label>	Branch if lower (C flag set)	BRBS 0, <label>
BRNE <label>	Branch if not equal (Z flag clear)	BRBC 1, <label>
BREQ <label>	Branch if equal (Z flag set)	BRBS 1, <label>
BRPL <label>	Branch if plus (N flag clear)	BRBC 2, <label>
BRMI <label>	Branch if minus (N flag set)	BRBS 2, <label>
BRVC <label>	Branch to <label> if V flag is clear	BRBC 3, <label>
BRVS <label>	Branch to <label> if V flag is set	BRBS 3, <label>
BRLT <label>	Branch if less than (S flag clear)	BRBC 4, <label>
BRGE <label>	Branch if greater or equal (S flag set)	BRBS 4, <label>
BRHC <label>	Branch to <label> if H flag is clear	BRBC 5, <label>
BRHS <label>	Branch to <label> if H flag is set	BRBS 5, <label>
BRTC <label>	Branch to <label> if T flag is clear	BRBC 6, <label>
BRTS <label>	Branch to <label> if T flag is set	BRBS 6, <label>
BRID <label>	Branch to <label> if interrupts disabled	BRBC 7, <label>
BRIE <label>	Branch to <label> if interrupts enabled	BRBS 7, <label>
CLC	Clears Carry Flag	BCLR 0
CLZ	Clears Zero Flag	BCLR 1
CLN	Clears Negative Flag	BCLR 2
CLV	Clears V (two's complement) Flag	BCLR 3
CLS	Clears Sign Flag	BCLR 4
CLH	Clears Half Carry Flag	BCLR 5
CLT	Clears Temp Flag	BCLR 6
CLI	Clears I bit (disables interrupts)	BCLR 7
SEC	Sets Carry Flag	BSET 0
SEZ	Sets Zero Flag	BSET 1
SEN	Sets Negative Flag	BSET 2
SEV	Sets V (two's complement) Flag	BSET 3
SES	Sets Sign Flag	BSET 4
SEH	Sets Half Carry Flag	BSET 5
SET	Sets Temp Flag	BSET 6
SEI	Sets I bit (enables interrupts)	BSET 7
	Shaded instructions refer to instructions useful after compare or subtract instructions, such as CP, CPI, SUB and SUBI.	

Appendix D
Instruction glossary

Here is a list of all instructions used by the standard and Tiny AVRs. The Mega AVRs have a few more instructions (involving, for example, multiplication).

The following names are used in the descriptions:

reg	refers to:	any of the 32 working registers
hreg	refers to:	the higher half of the working registers (16–31)
ioreg	refers to:	any of the 64 input/output registers
lioreg	refers to:	the lower half of the I/O registers (0–31)
longreg	refers to:	one of the 16-bit 'long' registers (e.g. X, Y, Z)

adc   reg1, reg2   [HSVNZC]
- **add**s the number in **reg1**, the number in **reg2**, and the carry bit leaving the result in reg1

add   reg1, reg2   [HSVNZC]
- **add**s the number in **reg1** with the number in **reg2**, leaving the result in reg1

adiw   longreg, number   [SVNZC]
- *(Not for 1200 and Tiny AVRs)* – **ad**ds a **number** between 0 and 63 to one of the 16-bit 'long' registers (X, Y, Z)

and   reg1, reg2   [SVNZ]
- **AND**s the number in **reg1** with the number in **reg2**, leaving the result in reg1

andi   hreg, number   [SVNZ]
- **AND**s a **number** (0–255) with the number in an upper-half register, leaving the result in that register

asr   reg   [SVNZC]
- '**arithmetically**' shifts all the bits in **reg** to the right (bit 7 remains unchanged)

bclr   bit   [ITHSVNZC]
- **cl**ears a **bit** in SREG (i.e. makes it 0)

bld   reg, bit   [-]
- loads the T bit into a certain bit in a register

brbc   bit, label   [-]
- tests a bit in SREG, and branches (jumps) to **label** if the bit is clear. Note: the label must be within 63 instructions of the brbc instruction.

brbs   bit, label   [-]
- tests a bit in SREG, and branches (jumps) to **label** if the bit is set. Note: the label must be within 63 instructions of the brbs instruction

brcc	**label**	- tests Carry Flag, branches if clear
brcs	**label**	- tests Carry Flag, branches if set
breq	**label**	- tests Zero Flag, branches if set (regs are equal)
brge	**label**	- tests Sign Flag, branches if clear (greater or equal)
brhc	**label**	- tests Half Carry Flag, branches if clear
brhs	**label**	- tests Half Carry Flag, branches if set
brid	**label**	- tests Interrupt Flag, branches if clear (disabled)
brie	**label**	- tests Interrupt Flag, branches if set (enabled)
brlo	**label**	- tests Carry Flag, branches if set (lower)
brlt	**label**	- tests Sign Flag, branches if set (less than)
brmi	**label**	- tests Negative Flag, branches if set (minus)
brne	**label**	- tests Zero Flag, branches if clear (regs are not equal)
brpl	**label**	- tests Negative Flag, branches if clear (plus)
brsh	**label**	- tests Carry Flag, branches if set (same or higher)
brtc	**label**	- tests T Flag, branches if clear
brts	**label**	- tests T Flag, branches if set
brvc	**label**	- tests Overflow Flag, branches if clear
brvs	**label**	- tests Overflow Flag, branches if set

bset   bit   [ITHSVNZC]
- sets a bit in SREG (i.e. makes it 1)

bst   reg, bit   [T]
- stores a certain bit in a register in the T bit

call   label   [-]
- *(Only for Mega AVRs)* – calls the subroutine given by **label**, which can be anywhere in the program

cbi   **lioreg, bit**   [-]
- clears (makes 0) a bit in one of the lower-half I/O registers (0–31)

cbr   **reg, binary**   [SVNZ]
- clears some bits in a register, according to the 8-bit **binary** number in which a 0 means 'clear this bit' and a 1 means 'leave this bit alone'

clc	- clears Carry Flag	[C]
clh	- clears Half Carry Flag	[H]
cli	- clears Interrupt Flag	[I]
cln	- clears Negative Flag	[N]

clr   **reg**   [SVNZ]
- clears a register (moves 0 into it)

cls	- clears Sign Flag	[S]
clt	- clears T Bit	[T]
clv	- clears Overflow Flag	[V]
clz	- clears Zero Flag	[Z]

com   **reg**   [SVNZC]
- complements a register (inverts all the bits – ones become zeros, zeros become ones)

cp   **reg1, reg2**   [HSVNZC]
- compares the numbers in **reg1** and **reg2**, effectively subtracts reg2 from reg1, whilst leaving both registers unchanged

cpc   **reg1, reg2**   [HSVNZC]
- compares the numbers in **reg1** and **reg2** taking into account the carry flag, effectively performs (reg1 minus reg2 minus carry flag), whilst leaving both registers unchanged

cpi   **hreg, number**   [HSVNZC]
- compares the number in **hreg** with a **number**, effectively subtracts number from reg1, whilst leaving the register unchanged

cpse   **reg1, reg2**   [-]
- compares the numbers in **reg1** and **reg2**, skipping the next instruction if they are equal

dec   **reg**   [SVNZ]
- decrements (subtracts one from) a register, leaving the result in the register

eor   reg1, reg2   [SVNZ]
- exclusive **OR**s the number in **reg1** with the number in **reg2**, leaving the result in reg1

icall   [-]
- *(Not for 1200 and Tiny AVRs)* – (indirectly) **call**s a subroutine with address given by Z

ijmp   [-]
- *(Not for 1200 and Tiny AVRs)* – (indirectly) **jump**s to the address given by Z

in   reg, ioreg   [-]
- copies the number in an I/O register **in**to a working register

inc   reg   [SVNZ]
- **inc**rements (adds one to) a register, leaving the result in the register

jmp   label   [-]
- *(Only for Mega AVRs)* – **jump**s to the section called by **label**, which can be anywhere in the program

ld   reg, longreg   [-]
- **lo**ads the memory location pointed to by **longreg** into a register (**reg**)

ld   reg, longreg+   [-]
- *(Not for 1200 and Tiny AVRs)* – **lo**ads the memory location pointed to by **longreg** into **reg**, and then adds one to longreg

ld   reg, -longreg   [-]
- *(Not for 1200 and Tiny AVRs)* – subtracts one from longreg, and then **lo**ads the memory location pointed to by **longreg** into **reg**

ldd   reg, longreg+number   [-]
- *(Not for 1200 and Tiny AVRs)* – **lo**ads the memory location pointed to by the **Y** or **Z** registers into **reg**, and then adds a **number** (0–63) to longreg *(Note: doesn't work with X)*

ldi   hreg, number   [-]
- **lo**ads a **number** (0-255) into an upper-half register (16–31)

lds   reg, number   [-]
- *(Not for 1200 and Tiny AVRs)* – **lo**ads the contents of memory at address (**number**) registers into **reg**, where number can be between 0 and 65 535 (i.e. up to 64K)

lpm [-]
- *(Not for 1200)* – loads into R0 the contents of the program memory at the
address specified by the Z register

lsl   reg [SVNZC]
- 'logically' shifts all the bits in **reg** to the left (bit 7 goes into Carry flag, bit 0
is 0)

lsr   reg [SVNZC]
- 'logically' shifts all the bits in **reg** to the right (bit 0 goes into Carry flag, bit
7 is 0)

mov   reg1, reg2 [-]
- copies (**mov**es) the number in **reg2** into **reg1**

neg   reg [HSVNZC]
- makes the number in a register **neg**ative (20 becomes -20, equivalent to 236)

nop [-]
- this stands for **n**o **op**eration, literally 'do nothing' – good for wasting a clock
cycle

or   reg1, reg2 [SVNZ]
- inclusive **OR**s the number in **reg1** with the number in **reg2**, leaving the result
in reg1

ori   hreg, number [SVNZ]
- inclusive **OR**s a **number** (0–255) with the number in an upper-half register,
leaving the result in that register

out   ioreg, reg [-]
- copies the number in a working register **out** to an I/O register

pop   reg [-]
- *(Not for 1200 and Tiny AVRs)* – **pop**s the top of the stack into a register

push   reg [-]
- *(Not for 1200 and Tiny AVRs)* – **push**es the contents of a register onto the stack

rcall   label [-]
- **call**s the subroutine labeled by **label**, which must be not further than 2048
instructions from the rjmp instructions (i.e. a relative call)

ret   [-]
- **ret**urns from a subroutine (executes the line after the original call instruction)

reti   [I]
- **ret**urns from a subroutine, and sets the interrupt flag

rjmp   **label**   [-]
- **jump**s to a part of the program labeled by **label**, which must be not further than 2048 instructions from the rjmp instructions (i.e a relative jump).

rol   **reg**   [SVNZC]
- **ro**tates all the bits in **reg** to the left (C flag goes into bit 0, bit 7 goes into C Flag)

ror   **reg**   [SVNZC]
- **ro**tates all the bits in **reg** to the right (C flag goes into bit 7, bit 0 goes into C flag)

sbc   **reg1, reg2**   [HSVNZC]
- **sub**tracts the number in **reg2** and the carry bit from **reg1** , leaving the result in reg1

sbci   **hreg, number**   [HSVNZC]
- **sub**tracts a **number** (0–255) and the carry bit from the number in an upper-half register, leaving the result in that register

sbi   **lioreg, bit**   [-]
- **s**ets (makes 1) a **b**it in one of the lower-half I/O registers (0–31)

sbic   **lioreg, bit**   [-]
- tests a **b**it in a lower-half I/O register (0–31), and **s**kips the next instruction if is **c**lear

sbis   **lioreg, bit**   [-]
- tests a **b**it in a lower-half I/O register (0–31), and **s**kips the next instruction if is **s**et

sbiw   **longreg, number**   [SVNZC]
- *(Not for 1200 and Tiny AVRs)* - **s**ubtracts a **number** between 0 and 63 from one of the 16-bit 'long' registers (X, Y, Z)

sbr   **reg, binary**   [SVNZ]
- **s**ets some **b**its in a **r**egister, according to the 8-bit **binary** number in which a 1 means 'set this bit' and a 0 means 'leave this bit alone'

sbrc   reg, bit   [-]
- tests a **bit** in a **reg**ister, and **s**kips the next instruction if is **c**lear

sbrs   reg, bit   [-]
- tests a **bit** in a **reg**ister, and **s**kips the next instruction if is **s**et

sec	- **se**ts **C**arry Flag	[C]
seh	- **se**ts **H**alf Carry Flag	[H]
sei	- **se**ts **I**nterrupt Flag	[I]
sen	- **se**ts **N**egative Flag	[N]

ser   reg   [-]
- **se**ts a **r**egister (moves 255 / $FF / 0b11111111 into it)

ses	- **se**ts **S**ign Flag	[S]
set	- **se**ts **T** Bit	[T]
sev	- **se**ts **O**verflow Flag	[V]
sez	- **se**ts **Z**ero Flag	[Z]

sleep   [-]
- sends the chip to **sleep**, a low-power mode (woken up through reset or interrupt)

st   reg, longreg   [-]
- **st**ores the number in a register (**reg**) to the memory location pointed to by **longreg**

st   reg, longreg+   [-]
- *(Not for 1200 and Tiny AVRs)* - **st**ores the number in **reg** to the memory location pointed to by **longreg**, and then adds one to longreg

st   reg, -longreg   [-]
- *(Not for 1200 and Tiny AVRs)* - subtracts one from longreg, and then **st**ores the number that's in **reg** to the memory location pointed to by **longreg**

std   reg, longreg+number   [-]
- *(Not for 1200 and Tiny AVRs)* - **st**ores the number in **reg** to the memory location pointed to by the **Y** or **Z** registers, and then adds a **number** (0–63) to longreg *(Note: doesn't work with X)*

sts   reg, number   [-]
- *(Not for 1200 and Tiny AVRs)* - **st**ores the number in **reg** at memory address (**number**), where number can be between 0 and 65535 (i.e. up to 64K)

sub   **reg1, reg2**   [HSVNZC]
- **sub**tracts the number in **reg2** from the number in **reg1**, leaving the result in reg1

subi   **hreg, number**   [HSVNZC]
- **sub**tracts a **number** (0–255) from the number in an upper-half register, leaving the result in that register

swap   **reg**   [SVNZC]
- **swap**s the upper and lower nibbles of a register, leaving the result in the register

tst   **reg**   [SVNZ]
- **tst**s to see if the number in a register is 0 by ANDing it with itself (leaving the register unchanged). The zero flag must then be tested using breq or brne to complete the test

wdr   [-]
- **r**esets the **w**atch**d**og timer (must be done at regular intervals to avoid reset)

Appendix G
ASCII conversion

	0	1	2	3	4	5	6	7	8	9	A	B	C	D	E	F
0	NUL	SOH	STX	ETX	EOT	ENQ	ACK	BEL	BS	TAB	LF	VT	FF	CR	SO	SI
1	DLE	DC1	DC2	DC3	DC4	NAK	SYN	ETB	CAN	EM	SUB	ESC	FS	GS	RS	US
2	SP	!	"	#	$	%	&	'	(	)	*	+	´	-	.	/
3	0	1	2	3	4	5	6	7	8	9	:	;	<	=	>	?
4	@	A	B	C	D	E	F	G	H	I	J	K	L	M	N	O
5	P	Q	R	S	T	U	V	W	X	Y	Z	[	\	]	^	_
6	`	a	b	c	d	e	f	g	h	I	j	k	l	m	n	o
7	p	q	r	s	t	u	v	w	x	y	z	{			}	~	DEL

e.g. 'T' = 0x54

NUL ($00):	null		DLE ($10):	data link escape
SOH ($01):	start of heading		DC1 ($11):	device control 1
STX ($02):	start of text		DC2 ($12):	device control 2
ETX ($03):	end of text		DC3 ($13):	device control 3
EOT ($04):	end of transmission		DC4 ($14):	device control 4
ENQ ($05):	enquiry		NAK ($15):	negative acknowledge
ACK ($06):	acknowledge		SYN ($16):	synchronous idle
BEL ($07):	bell		ETB ($17):	end of transmission block
BS ($08):	backspace		CAN ($18):	cancel
TAB ($09):	horizontal tab		EM ($19):	end of medium
LF ($0A):	line feed		SUB ($1A):	substitute
NL ($0A):	new line		ESC ($1B):	escape
VT ($0B):	vertical tab		FS ($1C):	file separator
FF ($0C):	form feed		GS ($1D):	group separator
NP ($0C):	new page		RS ($1E):	record separator
CR ($0D):	carriage return		US ($1F):	unit separator
SO ($0E):	shift out			
SI ($0F):	shift in			

Appendix H
When all else fails, read this

You should find that there are certain mistakes which you make time and time again (I do!). I've listed some popular ones here :

?   Have you put a colon after your labels, i.e. **start:** and *not* **start**?
?   Have you tried to use **sbi**, **cbi**, **sbis** or **sbic** with I/O registers $20–$3F?
?   Are you remembering to reset counting registers?
?   Have you set registers to correct initial values in Init?
?   Have you remembered that on Tiny10 and Tiny11, PB5 is input only?
?   Have you set up the stack pointer (SPL/SPH) if necessary?
?   Are you writing/reading 2-byte registers such as **TCNT1H,L** in the correct order?
?   If you are having a total nightmare and NOTHING is working ... have you specified the correct AVR at the top?

Appendix E
Interupt vector tables

	Tiny10 Tiny11	Tiny12	Tiny15	'1200	'2313	'2323 '2343	'2333 '4433	'4434 '8535	'8515	
$000	RESET	RESET	RESET	RESET	RESET	RESET	RESET	RESET	RESET	$000
$001	INT0	INT0	INT0	INT0	INT0	INT0	INT0	INT0	INT0	$001
$002	PinChange	PinChange	PinChange	T/C0 OVF	INT1	T/C0 OVF	INT1	INT1	INT1	$002
$003	T/C0 OVF	T/C0 OVF	T/C1 Comp	An. Comp.	T/C1 Capt.		T/C1 Capt.	TC2 Comp.	TC1 Capt	$003
$004	An. Comp.	EE. Ready	T/C1 OVF		T/C1 Comp		T/C1 Comp	TC2 OVF	TC1 CompA	$004
$005		An. Comp.	T/C0 OVF		T/C1 OVF		T/C1 OVF	TC1 Capt	TC1 CompB	$005
$006			EE. Ready		T/C0 OVF		T/C0 OVF	TC1 CompA	TC1 OVF	$006
$007			An. Comp.		UART Rx		Serial Done	TC1 CompB	TC0 OVF	$007
$008			ADC Done		UART Em.		UART Rx	TC1 OVF	Serial Done	$008
$009					UART Tx		UART Em.	TC0 OVF	UART Rx	$009
$00A					An. Comp.		UART Tx	Serial Done	UART Em.	$00A
$00B							ADC Done	UART Rx	UART Tx	$00B
$00C							EE Ready	UART Em.	An. Comp.	$00C
$00D							An. Comp.	UART Tx		$00D
$00E								ADC Done		$00E
$00F								EE Ready		$00F
$010								An. Comp.		$010

Appendix F
Hex conversion

	0	1	2	3	4	5	6	7	8	9	A	B	C	D	E	F
0	0	1	2	3	4	5	6	7	8	9	10	11	12	13	14	15
1	16	17	18	19	20	21	22	23	24	25	26	27	28	29	30	31
2	32	33	34	35	36	37	38	39	40	41	42	43	44	45	46	47
3	48	49	50	51	52	53	54	55	56	57	58	59	60	61	62	63
4	64	65	66	67	68	69	70	71	72	73	74	75	76	77	78	79
5	80	81	82	83	84	85	86	87	88	89	90	91	92	93	94	95
6	96	97	98	99	100	101	102	103	104	105	106	107	108	109	110	111
7	112	113	114	115	116	117	118	119	120	121	122	123	124	125	126	127
8	128	129	130	131	132	133	134	135	136	137	138	139	140	141	142	143
9	144	145	146	147	148	149	150	151	152	153	154	155	156	157	158	159
A	160	161	162	163	164	165	166	167	168	169	170	171	172	173	174	175
B	176	177	178	179	180	181	182	183	184	185	186	187	188	189	190	191
C	192	193	194	195	196	197	198	199	200	201	202	203	204	205	206	207
D	208	209	210	211	212	213	214	215	216	217	218	219	220	221	222	223
E	224	225	226	227	228	229	230	231	232	233	234	235	236	237	238	239
F	240	241	242	243	244	245	246	247	248	249	250	251	252	253	254	255

Appendix I
Contacts and further reading

John Morton: help@to-pic.com

ATMEL website: http://www.atmel.com

Kühnel, Claus (1998) *AVR RISC Microcontroller Handbook,* Newnes
(gives more details of the inner architecture of AVRs)

Brimicombe, M.W. (1985) *Electronic Systems*, Nelson
(great text for general electronics)

Some fun AVR projects: http://www.riccibitti.com/designs.htm

Random numbers: http://www.physics.carleton.ca/courses/75.502/slides/
monte12/

Appendix J
Sample programs

Program A - LEDon

```
;**********************************
; written by: John Morton          *
; date:   5/2/2002                 *
; version: 1.0                     *
; file saved as: LEDon.asm         *
; for AVR: 1200                    *
; clock frequency: 4MHz            *
;**********************************

; Program Function: Turns an LED on

.device    at90s1200
.nolist
.include    "C:\Program Files\Atmel\AVR Studio\Appnotes\1200def.inc"
.list


;==========
; Declarations:

.def  temp      =r16

;=============
; Start of Program

      rjmp   Init                 ; first line executed

;==========
Init: ser    temp                 ; PB0 - output, rest N/C
      out    DDRB, temp           ;
      out    DDRD, temp           ; PD0-7 all N/C

clr          temp                 ; all Port B outputs off
      out    PortB, temp          ;
      out    PortD, temp          ; all Port D N/C
```

```
;====================
Start:
    sbi   PortB, 0              ; turns on LED
    rjmp Start                  ; loops back to Start
```

Program B - Push Button

```
;***********************************
; written by: John Morton          *
; date:   5/2/2002                 *
; version: 1.0                     *
; file saved as: PushA.asm         *
; for AVR: 1200                    *
; clock frequency: 4MHz            *
;***********************************

; Program Function: Turns an LED on when a button is pressed

.device    at90s1200
.nolist
.include    "C:\Program Files\Atmel\AVR Studio\Appnotes\1200def.inc"
.list


;==========
; Declarations:

.def   temp       =r16


;=============
; Start of Program

      rjmp   Init                ; first line executed


;==========
Init: ser    temp               ; PB0 - output, rest N/C
      out    DDRB, temp         ;
      ldi    temp, 0b11111110   ; PD0 - input, rest N/C
      out    DDRD, temp         ;

      clr    temp               ; all Port B outputs off
      out    PortB, temp        ;
      ldi    temp, 0b00000001   ; PD0 - pull-up, rest N/C
      out    PortD, temp        ;
```

```
;====================
Start:
    sbis   PinD, 0          ; tests push button
    rjmp   LEDoff           ; goes to LEDoff
    sbi    PortB, 0         ; turns on LED
    rjmp   Start            ; loops back to Start

LEDoff:
    cbi    PortB, 0         ; turns off LED
    rjmp   Start            ; loops back to start
```

Program C - Push Button

```
;***********************************
; written by: John Morton          *
; date:   5/2/2002                 *
; version: 2.0                      *
; file saved as: PushB.asm          *
; for AVR: 1200                      *
; clock frequency: 4MHz             *
;***********************************

; Program Function: Turns an LED on when a button is pressed

.device    at90s1200
.nolist
.include   "C:\Program Files\Atmel\AVR Studio\Appnotes\1200def.inc"
.list


;==========
; Declarations:

.def   temp      =r16


;=============
; Start of Program

    rjmp   Init                 ; first line executed

;==========
Init: ser   temp               ; PB0 - output, rest N/C
      out   DDRB, temp         ;
      ldi   temp, 0b11111110   ; PD0 - input, rest N/C
      out   DDRD, temp         ;
```

```
clr       temp              ; all Port B outputs off
    out   PortB, temp       ;
    ldi   temp, 0b00000001  ; PD0 - pull-up, rest N/C
    out   PortD, temp       ;
```

```
;==================
Start:
    in   temp, PinD         ; reads button
    out  PortB, temp        ; controls LED
    rjmp Start              ; loops back
```

Program D - Counter

```
;************************************
; written by: John Morton          *
; date:   7/2/2002                 *
; version: 1.0                     *
; file saved as: counter.asm       *
; for AVR: 1200                    *
; clock frequency: 4MHz            *
;************************************
```

```
; Program Function: Counts the number of times a button is pressed (0-9)
```

```
.device    at90s1200
.nolist
.include   "C:\Program Files\Atmel\AVR Studio\Appnotes\1200def.inc"
.list
```

```
;==========
; Declarations:
```

```
.def  temp     =r16
.def  Counter  =r17
```

```
;============
; Start of Program
```

```
    rjmp  Init               ; first line executed
```

```
;=========
Init: ser   temp             ; PB0-7: outputs
    out   DDRB, temp         ;
    ldi   temp, 0b11111110   ; PD0: input, rest N/C
    out   DDRD, temp         ;
```

```
        ldi     R20, 0b01111110     ; initial code for a 0
        out     PortB, temp         ;
        ldi     temp, 0b00000001    ; PD0  - pull-up, rest N/C
        out     PortD, temp         ;

        ldi     R21, 0b00110000     ; code for a 1
        ldi     R22, 0b01101101     ; code for a 2
        ldi     R23, 0b01111001     ; etc.
        ldi     R24, 0b00110011     ;
        ldi     R25, 0b01011011     ;
        ldi     R26, 0b01011111     ;
        ldi     R27, 0b01110000     ;
        ldi     R28, 0b01111111     ;
        ldi     R29, 0b01111011     ; code for a 9
        clr     Counter             ; Counter initially 0


;==================
Start:  sbic    PinD, 0             ; button pressed?
        rjmp    Start               ; no, so keeps looping
        inc     Counter             ; yes, so adds 1 to Counter

        cpi     Counter, 10         ; is Counter = 10?
        brne    PC+2                ; no, so skips
        clr     Counter             ; yes, so resets Counter

        ldi     ZL, 20              ; zeros ZL to R20
        add     ZL, Counter         ; adds Counter to ZL
        ld      temp, Z             ; reads Rx into temp
        out     PortB, temp         ; outputs temp to Port B
        rjmp    Start               ; loops back to Start
```

Program E - Counter v. 2.0

```
;***********************************
; written by: John Morton          *
; date:    7/2/2002                *
; version: 2.0                      *
; file saved as: counter.asm        *
; for AVR: 1200                     *
; clock frequency: 4MHz             *
;***********************************

; Program Function: Counts the number of times a button is pressed (0-9)

.device     at90s1200
```

```
.nolist
.include    "C:\Program Files\Atmel\AVR Studio\Appnotes\1200def.inc"
.list

;==========
; Declarations:

.def  temp      =r16
.def  Counter   =r17

;=============
; Start of Program

      rjmp  Init                ; first line executed

;==========
Init: ser    temp               ; PB0-7: outputs
      out    DDRB, temp         ;
      ldi    temp, 0b11111110   ; PD0: input, rest N/C
      out    DDRD, temp         ;

      ldi    R20, 0b01111110    ; initial code for a 0
      out    PortB, temp        ;
      ldi    temp, 0b00000001   ; PD0 - pull-up, rest N/C
      out    PortD, temp        ;

      ldi    R21, 0b00110000    ; code for a 1
      ldi    R22, 0b01101101    ; code for a 2
      ldi    R23, 0b01111001    ; etc.
      ldi    R24, 0b00110011    ;
      ldi    R25, 0b01011011    ;
      ldi    R26, 0b01011111    ;
      ldi    R27, 0b01110000    ;
      ldi    R28, 0b01111111    ;
      ldi    R29, 0b01111011    ; code for a 9
      clr    Counter            ; starts with a 0

;====================
Start:sbic  PinD, 0             ; button pressed?
      rjmp  Start               ; no, so keeps looping
      inc   Counter             ; yes, so adds 1 to Counter

      cpi   Counter, 10         ; is Counter = 10?
      brne  PC+2                ; no, so skips
```

```
        clr     Counter             ; yes, so resets Counter

        ldi     ZL, 20              ; zeros ZL to R20
        add     ZL, Counter         ; adds digit to ZL
        ld      temp, Z             ; reads Rx into temp
        out     PortB, temp         ; outputs temp to Port B

ReleaseWait:
        sbis    PinD, 0             ; button released?
        rjmp    ReleaseWait         ; no, so keeps looping
        rjmp    Start               ; yes, so loops back to start
```

Program F - Chaser

```
;***********************************
; written by: John Morton          *
; date:   7/2/2002                 *
; version: 1.0                     *
; file saved as: chaser.asm        *
; for AVR: 1200                    *
; clock frequency: 2.4576MHz       *
;***********************************

; Program Function: Chases a pattern of LEDs at varying speeds

.device     at90s1200
.nolist
.include    "C:\Program Files\Atmel\AVR Studio\Appnotes\1200def.inc"
.list


;==========
; Declarations:

.def   temp     =r16
.def   Mark240  =r17
.def   Counter  =r18
.def   Speed    =r19


;==============
; Start of Program

        rjmp    Init                ; first line executed


;==========
Init:   ser     temp                ; PB0-7: outputs
```

```
        out    DDRB, temp        ;
        ldi    temp, 0b11111100  ; PD0, 1 - input, rest N/C
        out    DDRD, temp        ;

        ldi    temp, 0b00000001  ; initially just PB0 on
        out    PortB, temp       ;
        ldi    temp, 0b00000011  ; PD0, 1 - pull-ups, rest N/C
        out    PortD, temp       ;

        ldi    temp, 0b00000101  ; sets up timer to count at CK/1024
        out    TCCR0, temp       ;

        ldi    Mark240, 240      ;
        ldi    Counter, 5        ;
        ldi    Speed, 5          ;

;====================
Start: sbic   PinD, 0            ; checks down button
        rjmp   UpTest            ; not pressed, jumps
        inc    Speed             ; slows down time
        cpi    Speed, 11         ; has Speed reached 11?
        brne   ReleaseDown       ; jumps to ReleaseDown if not equal
        dec    Speed             ; subtracts one from Speed
ReleaseDown:
        sbis   PinD, 0           ; waits for down button to be released
        rjmp   ReleaseDown       ;

UpTest:
        sbic   PinD, 1   ; checks up button
        rjmp   Timer             ; not pressed, jumps
        dec    Speed             ; speeds up time
        brne   ReleaseUp         ; jumps to Timer if not 0
        inc    Speed             ; adds one to Speed
ReleaseUp:
        sbis   PinD, 0           ; waits for up button to be released
        rjmp   ReleaseUp         ;

Timer:
        in     temp, TCNT0       ; reads Timer 0 into temp
        cp     temp, Mark240     ; compares temp with Mark240
        brne   Timer             ; if not equal loops back to Timer

        subi   Mark240, -240     ; adds 240 to Mark240
        dec    Counter           ; subtracts one from Counter
        brne   Start             ; if not zero loops back to Start
```

; set time has passed, rotates LEDs

```
    mov  Counter, Speed      ; resets Counter

    in   temp, PortB         ; reads in current state
    lsl  temp                ; rotates to the left
    brcc PC+2                ; checks Carry Flag, skip if clear
    ldi  temp, 0b00000001    ; resets to PB0 on, others off

    out  PortB, temp         ; outputs to PortB
    rjmp Start               ; loops back to Start
```

Program G - Counter v. 3.0
```
;************************************
; written by: John Morton          *
; date:   9/2/2002                 *
; version: 3.0                     *
; file saved as: counter.asm       *
; for AVR: 1200                    *
; clock frequency: 4MHz            *
;************************************
```

; Program Function: Counts the number of times a button is pressed (0-9)

```
.device    at90s1200
.nolist
.include   "C:\Program Files\Atmel\AVR Studio\Appnotes\1200def.inc"
.list


;===========
; Declarations:

.def  temp      =r16
.def  Counter   =r17
.def  Delay1    =r16
.def  Delay2    =r18
.def  Delay3    =r19


;===============
; Start of Program

    rjmp Init                ; first line executed
;===========
; Subroutines
```

```
Debounce:
     ldi     Delay1, 0x80        ; sets up counting registers
     ldi     Delay2, 0x38        ;
     ldi     Delay3, 0x01        ;

Loop:
     subi    Delay1, 1           ; inserts delay
     sbci    Delay2, 0           ;
     sbci    Delay3, 0           ;
     brcc    Loop                ;
     ret                         ; returns from subroutine


;=========
Init: ser   temp                ; PB0-7: outputs
     out     DDRB, temp          ;
     ldi     temp, 0b11111110    ; PD0: input, rest N/C
     out     DDRD, temp          ;

     ldi     R20, 0b01111110     ; initial code for a 0
     out     PortB, temp         ;
     ldi     temp, 0b00000001    ; PD0 - pull-up, rest N/C
     out     PortD, temp         ;

     ldi     R21, 0b00110000     ; code for a 1
     ldi     R22, 0b01101101     ; code for a 2
     ldi     R23, 0b01111001     ; etc.
     ldi     R24, 0b00110011     ;
     ldi     R25, 0b01011011     ;
     ldi     R26, 0b01011111     ;
     ldi     R27, 0b01110000     ;
     ldi     R28, 0b01111111     ;
     ldi     R29, 0b01111011     ; code for a 9
     clr     Counter             ; Counter initially 0


;====================
Start:sbic  PinD, 0             ; button pressed?
     rjmp    Start               ; no, so keeps looping
     inc     Counter             ; yes, so adds 1 to Counter

     cpi     Counter, 10         ; is Counter = 10?
     brne    PC+2                ; no, so skips
     clr     Counter             ; yes, so resets Counter

     ldi     ZL, 20              ; zeros ZL to R20
     add     ZL, digit           ; adds digit to ZL
```

```
      ld    temp, Z              ; reads Rx into temp
      out   PortB, temp          ; outputs temp to Port B

      rcall Debounce             ; inserts required delay
ReleaseWait:
      sbis  PinD, 0              ; button released?
      rjmp ReleaseWait           ; no, so keeps looping
      rcall Debounce             ; inserts required delay
      rjmp Start                 ; yes, so loops back to start
```

Program H - Traffic lights

```
;***********************************
; written by: John Morton                *
; date:    7/2/2002                      *
; version: 1.0                           *
; file saved as: traffic.asm             *
; for AVR: 1200                          *
; clock frequency: 2.4576MHz             *
;***********************************

; Program Function: Simulates a pedestrians crossing

.device     at90s1200
.nolist
.include    "C:\Program Files\Atmel\AVR Studio\Appnotes\1200def.inc"
.list


;==========
; Declarations:

.def  temp      =r16
.def  Counter   =r17
.def  tog       =r18
.def  Delay1    =r19
.def  Delay2    =r20
.def  Delay3    =r21
.def  Mark240   =r22
.def  Count250  =r23


;=============
; Start of Program

      rjmp  Init                 ; first line executed
```

```
;===========
; Subroutines:

HalfSecond:
        clr     Delay1          ; sets up counting registers
        ldi     Delay2, 0xC0    ;
        ldi     Delay3, 0x03    ;

HalfLoop:
        subi    Delay1, 1       ; inserts delay
        sbci    Delay2, 0       ;
        sbci    Delay3, 0       ;
        brcc    HalfLoop        ;
        ret                     ;

;=========
Timer:
        brts    PC+2            ; test T bit, skip if set
        ret                     ; returns if T is clear

        in      temp, TCNT0     ; reads Timer 0 into temp
        cpse    temp, Mark240   ; compares temp with Mark240
        ret                     ; if not equal returns
        subi    Mark240, -240   ; adds 240 to Mark240
        dec     Count250        ; subtracts one from Count250
        breq    PC+2            ; if zero, skips
        ret                     ; if not zero returns

        ldi     Count250, 250   ; resets Count250
        clt                     ; clears T bit
        ret                     ;

;=========
Init:   ser     temp            ; PB0-5: outputs, rest N/C
        out     DDRB, temp      ;
        ldi     temp, 0b11111110 ; PD0 - input, rest N/C
        out     DDRD, temp      ;

        ldi     temp, 0b00000001 ; PD0 - pull-up, rest N/C
        out     PortD, temp     ;

        ldi     temp, 0b00000101 ; sets up timer to count at
        out     TCCR0, temp     ;    CK/1024

        ldi     Mark240, 240    ;
```

```
        ldi     Count250, 250       ;
        clt                         ; clears T bit


;==================
Start:ldi       temp, 0b00010001    ; motorists: green
        out     PortB, temp         ; pedestrians: temp
        rcall   Timer               ; keeps timing
        sbic    PinD, 0             ; tests button
        rjmp    Start               ; not pressed

        sbi     PortB, 5            ; turns on WAIT light

Loop:
        rcall   Timer               ; keeps timing
        brts    Loop                ; stays in loop until T is clear
        sbi     PortB, 1            ; motor amber on
        cbi     PortB, 0            ; motor green off

        ldi     temp, 8             ; 4 second delay
FourSeconds:
        rcall   HalfSecond          ;
        dec     temp                ;
        brne    FourSeconds         ;

        ldi     temp, 0b00001100    ; motorists: red
        out     PortB, temp         ; pedestrians: green

        ldi     temp, 16            ; 8 second delay
EightSeconds:
        rcall   HalfSecond          ;
        dec     temp                ;
        brne    EightSeconds        ;

        ldi     tog, 0b00001010     ; motorists: amber
        out     PortB, tog          ; pedestrians: green

        ldi     Counter, 8          ; sets up Counter register

FlashLoop:
        rcall   HalfSecond          ; waits ½ a second
        in      temp, PinB          ; reads in state of lights
        eor     temp, tog           ; toggles
        out     PortB, temp         ; outputs
        dec     Counter             ; does this 8 times
        brne    FlashLoop           ;
```

```
        set                         ; sets T bit
        rjmp   Start                ; loops back to Start
```

Program I - Logic Gates
```
;*************************************
; written by: John Morton           *
; date:   9/2/2002                  *
; version: 1.0                      *
; file saved as: logic.asm          *
; for AVR: Tiny12                   *
; clock frequency: 2.4576MHz        *
;*************************************

; Program Function: Simulates AND, NAND, IOR, NOR, XOR, XNOR,
; NOT and buffer gates

.device     atTiny12
.nolist
.include    "C:\Program Files\Atmel\AVR Studio\Appnotes\tn12def.inc"
.list

;===========
; Declarations:

.def   temp        =r16

;=============
; Start of Program

        rjmp Init                   ; first line executed

;============
; Lookup Table:

.dw   0b0000000100010011            ; code for AND and IOR
.dw   0b0011001000100000            ; NAND and NOR
.dw   0b0010000100010010            ; ENOR and EOR
.dw   0b0011000000000011            ; NOT and buffer

;=========
Init: ldi     temp, 0b000001        ; PB0: output, rest inputs
      out     DDRB, temp            ;
      ldi     temp, 0b111110        ; PB1-5: pull-ups
```

```
       out    PortB, temp         ; PB0 initially off


;===================
Start:in      ZL, PinB            ; reads in PinB
       andi   ZL, 0b001110        ; masks 0, 4 and 5
       lsr    ZL                  ; rotates
       subi   ZL, -2              ; adds 2 to ZL
       lpm                        ; reads lookup table into R0

       sbis   PinB, 4             ; tests Input A
       swap   R0                  ; swaps nibbles if low
       sbis   PinB, 5             ; tests Input B
       ror    R0                  ; rotates right if low

       mov    temp, R0            ; copies R0 to temp
       ori    temp, 0b11110       ; forces bits 1-4 high
       out    PortB, temp         ; outputs result
       rjmp   Start               ; loops back to Start
```

Program J – Frequency Counter
```
;************************************
; written by: John Morton          *
; date:   14/02/02                 *
; version: 1.0                     *
; file saved as Frequency          *
; for AT90s8515                    *
; clock frequency: 4MHz            *
;************************************

; Program Function: To display the frequency of the input on 3 seven
; segment displays

.device at90s1200
.nolist
.include "C:\Program Files\Atmel\AVR Studio\Appnotes\1200def.inc"
.list


;==============
; Declarations

.def   temp      =r16
.def   temp2     =r17
.def   temp3     =r18
```

```
.def   lowerbyte          =r19
.def   upperbyte          =r20

.def   DisplayCounter     =r21
.def   DisplayNumber      =r22

.def   Delay1             =r23
.def   Delay2             =r24
.def   Delay3             =r25

.def   Hundreds           =r26
.def   Tens               =r27
.def   Ones               =r28

.def   store              =r29
.def   store2             =r19
.def   Counter            =r20


; R0-R12 are for display


;============
; Reset Table

       rjmp  Init                     ; calls initialization subroutine


;===============
; Initialization

Init: ser    temp              ; PB0 LED for Hz / kHz
      out    DDRB, temp        ; PB1-7 are seven segment display

      ldi    temp, 0b11101111  ; PD0-2 choose a display
      out    DDRD, temp        ; PD4 input, rest N/C

      clr    temp              ; no pull-ups
      out    PORTB, temp       ; all outputs off

      ldi    temp, 0b00000001  ; starts by selecting one
      out    PORTD, temp       ; all outputs off

      ldi    temp, 0b00001110  ; watchdog barks every second
      out    WDTCR, temp       ;
```

```
ldi     temp, 0b00110000        ; enables deep sleep function
out     MCUCR, temp             ;

ldi     Hundreds, 12
ldi     Tens, 12
ldi     Ones, 12
clr     ZH                      ; makes sure higher byte of Z is clear

ldi     DisplayCounter, 50      ;
clr     DisplayNumber

ldi     temp, 0b11111100        ; 0
mov     R0, temp
ldi     temp, 0b01100000        ; 1
mov     R1, temp
ldi     temp, 0b11011010        ; 2
mov     R2, temp
ldi     temp, 0b11110010        ; 3
mov     R3, temp
ldi     temp, 0b01100110        ; 4
mov     R4, temp
ldi     temp, 0b10110110        ; 5
mov     R5, temp
ldi     temp, 0b10111110        ; 6
mov     R6, temp
ldi     temp, 0b11100000        ; 7
mov     R7, temp
ldi     temp, 0b11111110        ; 8
mov     R8, temp
ldi     temp, 0b11110110        ; 9
mov     R9, temp
ldi     temp, 0b01101110        ; H
mov     R10, temp
ldi     temp, 0b00000010        ; -
mov     R11, temp
rjmp    Start
```

;===================
; Display Subroutine

Display:
```
dec     DisplayCounter          ; changes display every 50 visits
breq    PC+2                    ;
ret                             ;
```

```
        wdr                             ; pats the dog
        ldi     DisplayCounter, 50      ;

        inc     DisplayNumber           ;
        cpi     DisplayNumber,3         ;
        brne    PC+2                    ;
        clr     DisplayNumber           ;

        ldi     ZL, 26                  ; zeros ZL to R25
        add     ZL, DisplayNumber       ;
        ld      temp, Z                 ; copies number to convert into temp
        clr     ZL                      ; zeros ZL to R0
        add     ZL, temp                ; adds temp to ZL
        ld      temp, Z                 ; reads Rx into temp
        sbic    PortB, 7                ; tests kHz LED
        ori     temp, 0b10000000        ; if it's on, keeps it on

        out     PortB, temp             ; outputs temp to Port B

        in      temp, PinD              ;
        lsl     temp                    ;
        sbrc    temp, 3                 ;
        ldi     temp, 0b00000001        ;
        out     PortD, temp             ;

        ret                             ;
```

;==
; Converts 4 digit hex answer into three decimal digits

```
DigitConvert:
        clr     Hundreds                ;
        clr     Tens                    ;
        clr     Ones

FindHundreds:
        subi    lowerbyte, 100          ;
        sbci    upperbyte, 0            ;
        brcs    FindTens                ;
        inc     Hundreds                ;
        rjmp    FindHundreds            ;

FindTens:
        subi    lowerbyte, -100
```

```
        subi    lowerbyte, 10        ;
        brcs    FindOnes
        inc     Tens
        rjmp    FindTens+1

FindOnes:
        subi    lowerbyte, -10      ; adds back the last 10
        mov     ones, lowerbyte     ; number left in lowerbyte = ones
        ret                         ; finished
```

;==

; PROGRAM START

;==
; high speed counting for frequencies more than 1kHz

```
Start:ldi      Delay1, 00           ;
        ldi     Delay2, 0x7D         ;
        ldi     temp, 0b10000000    ; resets displays and turns on kHz LED
        out     PortB, temp          ;

        ldi     temp, 0b00000111    ; sets TCNT0 to count rising edge
        out     TCCR0, temp          ;    on T0 (PD4)
        clr     upperbyte
        out     TCNT0, upperbyte
        in      temp, TCNT0          ;

HighSpeed:
        subi    Delay1,1             ; counts for 0.064 seconds
        sbci    Delay2, 0            ;
        brcs    DoneHi               ;
        mov     temp2, temp          ;
        in      temp, TCNT0          ;
        cp      temp, temp2          ;
        brsh    HighSpeed            ; 8 cycles

        inc     upperbyte            ;
        cpi     upperbyte, 0xFA     ; too high?
        breq    TooHigh              ;
        subi    Delay1, 1            ;
        sbci    Delay2, 0            ;
        brcs    DoneHi               ;
```

```
        nop                     ;
        rjmp   HighSpeed        ;

DoneHi:
        in     lowerbyte, TCNT0 ; immediately stores TCNT0 value
        cp     lowerbyte, temp  ; compares with previous value
        brsh   PC+2             ;
        inc    upperbyte        ;
        cpi    upperbyte, 0xFA  ;
        breq   TooHigh          ;

Divide64:
        ldi    temp, 6          ;
        lsr    upperbyte        ;
        ror    lowerbyte        ;
        dec    temp             ;
        brne   Divide64+1       ;

        cpi    upperbyte,0      ; higher byte 0?
        brne   PC+3             ; skips next 2 instructions
        cpi    lowerbyte, 0     ; lower byte 0?
        breq   LowSpeed         ; if frequency less than 1kHz we should
                                ;     use lower frequency mode

        rcall  DigitConvert     ;

        ldi    Delay1, 0x2A     ;
        ldi    Delay2, 0xC6     ;
        ldi    Delay3, 0x01     ;
HalfSecond:
        rcall  Display          ; calls display for half a second
        subi   Delay1, 1        ;
        sbci   Delay2, 0        ;
        sbci   Delay3, 0        ;
        brcc   HalfSecond       ;
        rjmp   Start            ;

TooHigh:
        ldi    Hundreds, 11     ;
        ldi    Tens, 10         ;
        ldi    Ones, 1          ;
        rjmp   HalfSecond-3     ;
```

```
;================================================
; low speed counting for frequencies less than 1kHz

LowSpeed:
     ldi     temp, 0b00000001   ; sets TCNT0 to count at CK
     out     TCCR0, temp        ;
     clr     Delay2             ;
     clr     Delay3             ;
     cbi     PortB, 7           ; clears PortB, 7 to turn on Hz LED

     in      store, PinD        ; stores initial value
FirstChange:
     rcall   Display            ; keeps displays going
     in      store2, PinD       ;
     eor     store2, store      ; compares with current value
     andi    store2, 0b00010000 ; ignores all bits except PD4
     breq    FirstChange        ; keeps looping until PD4 changes

     ldi     Counter, 2         ; sets up Counter to 2
     clr     temp2              ; resets Timer0
     out     TCNT0, temp2       ;

     in      store, PinD        ; stores initial value
LowLoop:
     in      store2, PinD       ;
     eor     store2, store      ; compares with current value
     andi    store2, 0b00010000 ; ignores all bits except PD4
     brne    Change             ; jumps to Change if PD4 changes

     rcall   Display            ; keeps display going

     mov     temp2, Delay1      ;
     in      Delay1, TCNT0      ;
     cp      Delay1, temp2      ;
     brsh    LowLoop            ;
     inc     Delay2             ;
     brne    LowLoop            ;
     inc     Delay3             ;
     cpi     Delay3, 0x3E       ; too slow?
     breq    TooSlow            ;
     rjmp    LowLoop

Change:
     in      store, PortB       ; updates new value
     dec     Counter            ;
```

```
        brne  LowLoop

        ldi   temp, 0x0F
        ldi   temp2, 0x00
        cpi   Delay1, 0xA0      ;
        cpc   Delay2, temp      ;
        cpc   Delay3, temp2     ;
        brcs  Start                 ; yes, so goes to HighSpeed

        ldi   temp, 0x00
        ldi   temp2, 0x09
        ldi   temp3, 0x3D
        clr   lowerbyte
        clr   upperbyte

Divide:
        sub temp, Delay1
        sbc   temp2, Delay2
        sbc   temp3, Delay3
        brcs  DoneDividing
        inc   lowerbyte          ;
        brne  Divide
        inc   upperbyte
        rjmp  Divide

DoneDividing:
        rcall  DigitConvert
        rjmp  LowSpeed

TooSlow:
        out   PortD, temp        ; turns off Display
        sleep
        rjmp  LowSpeed
```

Program K – Reaction Tester

```
;************************************
; written by: John Morton            *
; date:   25/2/02                    *
; version: 1.0                       *
; file saved as: reaction.asm        *
; for AVR: 1200                      *
; clock frequency: 4MHz              *
;************************************

; Program Function: Reaction Tester
```

```
.device    at90s1200
.nolist
.include   "C:\Program Files\Atmel\AVR Studio\Appnotes\1200def.inc"
.list


;===========
; Declarations:

.def  temp           =r16
.def  Random         =r17
.def  Five           =r18
.def  TimeL          =r19
.def  TimeH          =r20
.def  Hundreds       =r21
.def  Tens           =r22
.def  Ones           =r23
.def  CountX         =r24
.def  DisplayNumber  =r25
.def  DisplayCounter =r26
.def  tempH          =r27
.def  Count4         =r28


;==============
; Start of Program

      rjmp  Init              ; first line executed
      rjmp  ExtInt            ;
      rjmp  TCNT0Int          ;


;==========================
ExtInt:
      sbis  PinD, 0           ; tests LED
      rjmp  Cheat             ;
      clr   temp              ; stops TCNT0
      out   TCCR0             ;
      in    TimeL, TCNT0      ; reads in TCNT0 value
      in    temp, TIFR        ; test for TCNT0 overflow
      sbrc  temp, 1           ;
      inc   TimeH             ;
      subi  TimeL, 0xA2       ; subtracts back 0xA2 from
      sbci  TimeH, 0          ;    total reaction time
      ldi   temp, 0b00000101  ; restarts TCNT0 at CK/1024
      out   TCCR0, temp       ;

      ldi   Count4, 4         ; Multiplies reaction time
```

```
        mov    temp, TimeL      ;    by five
        mov    tempH, TimeH     ;
Times5:
        add    temp, TimeL      ;
        adc    tempH, TimeH     ;
        dec    Count4           ;
        brne   Times5           ;

        clr    TimeL            ;
        clr    TimeH
Divide12:
        subi   temp, 12         ;
        sbci   tempH, 0         ;
        brcs   DoneDividing     ;
        inc    TimeL            ;
        brne   Divide12         ;
        inc    TimeH            ;
        rjmp   Divide12         ;

DoneDividing:
        rcall  DigitConvert     ;
        ret                     ; returns DOESN'T enable interrupts

Cheat:
        ldi    Hundreds, 10 ; b
        ldi    Tens, 11         ; A
        ldi    Ones, 12         ; d
        ret                     ; returns and DOESN'T enable interrupts

;========================
TCNT0Int:
        sbic   PinD, 0          ; tests LED
        rjmp   TInt_LEDon       ;
        dec    CountX           ;
        breq   PC+2             ;
        reti
        ldi    temp, 0xA2       ;
        out    TCNT0, temp      ;
        sbi    PortD, 0         ; turns on LED
        reti                    ;

TInt_Ledon:
        inc    TimeH            ; increments higher byte
        cpi    TimeH, 0x0A      ; tests for maximum time
        breq   PC+2             ; skips if too slow
```

```
        reti
        ldi     Hundreds, 13        ; -
        ldi     Tens, 14            ; H
        ldi     Ones, 1             ; I
        ret                         ;
```

;==================
; Display Subroutine

```
Display:
        dec     DisplayCounter      ; changes display every 50 visits
        breq    PC+2                ;
        ret                         ;
        wdr                         ; pats the dog
        ldi     DisplayCounter, 50  ;

        inc     DisplayNumber       ;
        cpi     DisplayNumber,3     ;
        brne    PC+2                ;
        clr     DisplayNumber       ;

        ldi     ZL, 21              ; zeros ZL to R21
        add     ZL, DisplayNumber   ;
        ld      temp, Z             ; copies number to convert into temp
        ldi     ZL, 0               ; zeros ZL to R0
        add     ZL, temp            ; adds temp to ZL
        ld      temp, Z             ; reads Rx into temp
        out     PortB, temp         ; outputs temp to Port B

        brtc    PC+2                ; tests T bit
        sbi     PortB, 0            ; turns on kHz LED

        in      temp, PinD          ;
        lsl     temp                ;
        brcs    PC+2                ;
        ldi     temp, 0b00100000    ;
        ori     temp, 0b00000110
        out     PortD, temp         ;

        ret                         ;
```

;==
; Converts 4 digit hex answer into three decimal digits
DigitConvert:

```
        clr     Hundreds        ;
        clr     Ones            ;
        clr     Tens            ;

FindHundreds:
        subi    TimeL, 100      ;
        sbci    TimeH, 0        ;
        brcs    FindTens        ;
        inc     Hundreds        ;
        rjmp    FindHundreds    ;

FindTens:
        subi    TimeL, -100
        subi    TimeL, 10       ;
        brcs    FindOnes
        inc     Tens
        rjmp    FindTens+1

FindOnes:
        subi    TimeL, -10      ; adds back the last 10
        mov     Ones, TimeL     ; number left in lowerbyte = ones
        ret                     ; finished


;========================
Init:   ldi     temp, 0b11111111 ; PB1-7: outputs, PB0: N/C
        out     DDRB, temp      ;
        ldi     temp, 0b11111001 ; PD0,4-6: outputs, PD3,7: N/C
        out     DDRD, temp      ; PD1,2: inputs

        ldi     temp, 0b00000000 ;
        out     PortB, temp     ;
        ldi     temp, 0b00100110 ; selects first display, pull-ups
        out     PortD, temp     ;    on both buttons

        ldi     temp, 0b00000101 ; TCNT0 at CK/1024
        out     TCCR0, temp     ;

        ldi     temp, 0b00000000 ; INT0 interrupt on falling edge
        out     MCUCR, temp     ;

        ldi     temp, 0b01000000 ; enables INT0 interrupt
        out     GIMSK, temp     ;
```

```
        ldi     temp, 0b00000010      ; enables TCNT0 interrupt
        out     TIMSK, temp           ;

        ldi     DisplayCounter, 50    ;
        clr     DisplayNumber

        ldi     temp, 0b11111100      ; 0
        mov     R0, temp
        ldi     temp, 0b01100000      ; 1
        mov     R1, temp
        ldi     temp, 0b11011010      ; 2
        mov     R2, temp
        ldi     temp, 0b11110010      ; 3
        mov     R3, temp
        ldi     temp, 0b01100110      ; 4
        mov     R4, temp
        ldi     temp, 0b10110110      ; 5
        mov     R5, temp
        ldi     temp, 0b10111110      ; 6
        mov     R6, temp
        ldi     temp, 0b11100000      ; 7
        mov     R7, temp
        ldi     temp, 0b11111110      ; 8
        mov     R8, temp
        ldi     temp, 0b11110110      ; 9
        mov     R9, temp
        ldi     temp, 0b00111110      ; b
        mov     R10, temp
        ldi     temp, 0b11101110      ; A
        mov     R11, temp
        ldi     temp, 0b01111010      ; d
        mov     R12, temp


;====================
; Main body of program:

Start:rcall   Display                 ; keeps display going
        sbic    PinD, 1               ; waits for Ready button
        rjmp    Start                 ; keeps looping until it's pressed

                                      ; gets next random number
        mov     temp, Random          ; multiplies by 5 and...
        add     Random, temp          ;
        add     Random, temp          ;
```

```
        add    Random, temp    ;
        add    Random, temp    ;
        inc    Random          ; ...adds 1

        mov    CountX, Random  ;
        lsr    CountX          ; divides by 2 and adds 60
        subi   CountX, -60     ;

        ldi    temp, 0b0100000 ; resets INT0 interrupt flag
        out    GIFR            ;
        ldi    temp, 0b00000010 ; resets TC0 overflow interrupt flag
        out    TIFR            ;
        sei                    ; enables interrupts
        clr    TimeH           ; reset time register
        out    PortB, TimeH    ; also turns off displays while waiting
Loopy:
        brid   Start           ; skips out when interrupts disabled
        rjmp   Loopy           ; Loops
```

Program L - 4-bit Analogue to Digital Converter

```
;************************************
; written by: John Morton          *
; date:   25/2/02                  *
; version: 1.0                     *
; file saved as: atod.asm          *
; for AVR: 1200                    *
; clock frequency: 4MHz            *
;************************************

; Program Function: 4-bit A-D converter

.device    at90s1200
.nolist
.include   "C:\Program Files\Atmel\AVR Studio\Appnotes\1200def.inc"
.list


;==========
; Declarations:

.def   temp     =r16


;=============
; Start of Program
```

```
        rjmp  Init                    ; first line executed

;==========================
Init: ldi    temp, 0b11111100   ; PB0,1: Analogue inputs
      out    DDRB, temp         ; PD2-7: N/C
      ldi    temp, 0b11111111   ; PD0-3: outputs, PD4-7: N/C
      out    DDRD, temp         ;

      clr    temp               ;
      out    PortB, temp        ;
      ldi    temp, 0b00001000   ; selects msb
      out    PortD, temp        ;

      ldi    temp, 0b10000000   ; turns on Analogue comparator
      out    ACSR, temp         ;

;==================
; Main body of program:

Start:sbis  ACSR, 5             ; checks AC result
      cbi    PortD, 3           ; clears bit 3
      sbi    PortD, 2           ;

      sbis   ACSR, 5            ; checks AC result
      cbi    PortD, 2           ; clears bit 2
      sbi    PortD, 1           ;

      sbis   ACSR, 5            ; checks AC result
      cbi    PortD, 1           ; clears bit 1
      sbi    PortD, 0           ;

      sbis   ACSR, 5            ; checks AC result
      cbi    PortD, 0           ; clears bit 0

      in     temp, PortD        ; read in final answer
      swap   temp               ; swap
      out    PortB, temp        ; outputs

      rjmp   Start              ; keeps looping until it's pressed
```

PROGRAM M – Voltage Inverter

```
;************************************
; written by: John Morton          *
; date:   25/2/02                  *
; version: 1.0                     *
; file saved as: inverter.asm      *
; for AVR: 1200                    *
; clock frequency: 4MHz            *
;************************************

; Program Function: Outputs 5 – (input voltage)

.device    atTiny15
.nolist
.include   "C:\Program Files\Atmel\AVR Studio\Appnotes\Tn15def.inc"
.list

;==========
; Declarations:

.def   temp      =r16
.def   tempH     =r17
.def   Desired   =r18
.def   Actual    =r19

;==============
; Start of Program

       rjmp   Init                 ; first line executed

;=========================
Init:  ldi    temp, 0b011100   ; PB0,1,5: Inputs
       out    DDRB, temp       ; PB2-4: N/C
       clr    temp             ; no pull-ups
       out    PortB, temp      ;

       ldi    temp, 0b11101011 ; enables ADC, clock = CK/8
       out    ADCSR, temp
       clr    temp             ; selects ADC0, VCC as reference
       out    ADMUX, temp      ; no left adjusted

;===================
; Main body of program:

Start:cbi     ADMUX, 0         ; selects ADC0 input
```

```
        sbi     ADCSR, ADSC         ; starts conversion
        sbic    ADCSR, ADSC
        rjmp    Start+2

        in      Desired, ADCH       ; reads in 8-bit ADC result
        com     Desired             ; takes 5 - answer

        sbi     ADMUX, 0            ; selects ADC1 input
        sbi     ADCSR, ADSC         ; starts conversion on output
Wait:   sbic    ADCSR, ADSC         ; waits until conversion has finished
        rjmp    Wait

        in      Actual, ADCH        ; reads in ADC result of actual output
        cp      Actual, Desired     ; compares actual with desired
        brlo    TooLow              ; too low?
        cp      Desired, Actual     ;
        brlo    TooHigh             ; too high?
        cbi     DDRB, 0             ; just right, so makes PB0 an input
        rjmp    Start               ; reads ADC0 input again

TooLow:
        sbi     DDRA, 0             ; too low so makes PB0 an output
        sbi     PortB, 0            ;     and sets it
        rjmp    Start               ; reads ADC0 input again

TooHigh:
        sbi     DDRB, 0             ; too high, so makes PB0 an output
        cbi     PortB, 0            ;     and clears it
        rjmp    Start               ; reads ADC0 input again
```

PROGRAM N – Melody Maker

```
;*************************************
; written by: John Morton            *
; date:    22/3/02                   *
; version: 1.0                       *
; file saved as: music.asm           *
; for AVR: 2313                      *
; clock frequency: 4MHz              *
;*************************************

; Program Function: Plays a melody stored in the EEPROM

.device     at90s2313
.nolist
```

```
.include    "C:\Program Files\Atmel\AVR Studio\Appnotes\2313def.inc"
.list


;==========
; Declarations:

.def  temp      =r16
.def  NoteL     =r19
.def  NoteH     =r23
.def  Length    =r20
.def  address   =r21


;==============
; Start of Program

        rjmp  Init               ; first line executed
        reti                     ; $001 - INT0
        reti                     ; $002
        reti                     ; $003
        rjmp  ToggleOut          ; $004 - Compare A
        reti                     ; $005 - TC1 Overflow
        rjmp  ChangeNote         ; $006 - TC0 Overflow


;=====================
.org  0x13

LookUpTable:
.dw   0x0ECB        ;$00 = C
.dw   0x0DF7        ;$01 = C#
.dw   0x0D2E        ;$02 = D
.dw   0x0C71        ;$03 = D#
.dw   0x0BBE        ;$04 = E
.dw   0x0B15        ;$05 = F
.dw   0x0A76        ;$06 = F#
.dw   0x09E0        ;$07 = G
.dw   0x0952        ;$08 = G#
.dw   0x08CC        ;$09 = A
.dw   0x084D        ;$0A = A#
.dw   0x07D6        ;$0B = B


;=====================
ToggleOut:
      in    temp, PortD          ; toggles state of speaker output
      com   temp                 ; to produce square wave
```

```
        out    PortD, temp          ;
        reti


;====================
ChangeNote:
        dec    Length               ; waits sufficient length
        breq   PC+2                 ;
        reti                        ;


Rest: in       temp, TIFR          ; creates short pause between notes
      sbrs     temp, 1             ;
      rjmp     Rest                ;
      ldi      temp, 0b00000010 ;
      out      TIFR, temp          ;


;=============
Read_EEPROM:
        out    EEARL, address      ; reads next address
        sbi    EECR, 0             ; initiate read
                                   ; get note
        in     ZL, EEDR            ; reads EEPROM
        andi   ZL, 0b00001111      ; masks bits 4-7

        cpi    ZL, 0x0C            ; if 0x0C, loops back to first address
        breq   Reset               ;
        brlo   PC+2                ; if higher than 0C, makes 0B
        ldi    ZL, 0x0B            ;

        lsl    ZL                  ; multiplies by 2 to get word address
        subi   ZL, -0x26           ; adds 26 to point to table
        lpm                        ; reads look-up table
        mov    NoteL, R0           ; stores result
        inc    ZL                  ; reads next entry of look-up table
        lpm
        mov    NoteH, R0           ; stores result

                                   ; get octave
        in     temp, EEDR          ; reads EEPROM again
        swap   temp                ;
        andi   temp, 0b00000011 ; gets bits 4 and 5
GetOctave:
        breq   GetLength           ; uses bits 4,5 to select octave
        lsl    NoteL               ; divides by two to get next octave
        rol    NoteH               ;
```

```
        dec    temp            ;
        rjmp   GetOctave       ;

GetLength:                     ; gets length
        out    OCR1AH, NoteH   ; stores final freq values in Output
        out    OCR1AL, NoteL   ;    Compare registers
        in     temp, EEDR      ; reads EEPROM again
        andi   temp, 0b11000000 ; gets bits 6 and 7
        swap   temp            ;
        lsr    temp            ; uses these to get a Length = 2, 4, 6 or 8
        subi   temp, -2        ;
        mov    Length, temp    ; stores in Length
        inc    address         ; selects next EERPOM address (next note)
        reti

Reset:
        clr    address         ; resets EEPROM address to 0
        rjmp   Read_EEPROM     ;

;========================
Init: ldi    temp, 0b01000000 ; PB0-5: keyboard in
        out    DDRB, temp      ; PB6: N/C, PB7: Record
        ldi    temp, 0b01111011 ; PD0: N/C, PD1: speaker
        out    DDRD, temp      ; PD2: play, PD3-6: keyboard out

        ldi    temp, 0b10000000 ; no pull-ups on  PortB
        out    PortB, temp     ;
        ldi    temp, 0b00000100 ; pull-ups on play button
        out    PortD, temp     ;

        ldi    temp, 0b00000101 ; TC0 is CK/1024
        out    TCCR0, temp     ;
        clr    temp            ; no PWM
        out    TCCR1A, temp    ;
        ldi    temp, 0b00001001 ; TC1 is CK, clear TC1
        out    TCCR1B, temp    ;  after comparematch

        ldi    temp, 0b01000010 ; enables TC0 interrupt
        out    TIMSK, temp     ; enables TC1 CompA int.
        ldi    temp, 0b00000000 ; disables other interrupts
        out    GIMSK, temp     ;

        ldi    temp, RAMEND    ; sets up stack pointers
        out    SPL, temp
```

```
      clr    ZH              ;
      clr    address         ;
      out    EEARH, address  ;

      rcall  Read_EEPROM     ; gets first note
```

```
;===================
; Main body of program:

Start:rjmp  Start
```

PROGRAM O – Keyboard Converter
```
;************************************
; written by: John Morton            *
; date:    25/2/02                   *
; version: 1.0                       *
; file saved as: Keyboard.asm        *
; for AVR: 2313                      *
; clock frequency: 4MHz              *
;************************************

; Program Function: Converts a computer keyboard into a musical one

.device    at90s2313
.nolist
.include   "C:\Program Files\Atmel\AVR Studio\Appnotes\2313def.inc"
.list
```

```
;==========
; Declarations:

.def  temp      =r16
.def  data      =r17
.def  Length    =r18
```

```
;=================
; Start of Program

rjmp Init               ; first line executed
     reti               ; $001 - INT0
     reti               ; $002
     reti               ; $003
     reti               ; $004 - Compare A
     reti               ; $005
```

```
        reti                    ; $006 - TC1 Overflow
        rjmp    EndNote         ; $007 - TC0 Overflow
        reti                    ; $008
        rjmp    Change          ; $009 - UART Received
        reti                    ; $00A
        reti                    ; $00B
        reti                    ; $00C
```

```
;=====================
.org    13
;Note Lookup Table

.dw     0x1E84                  ;'a' = C
.dw     0xFFFF, 0xFFFF          ;'b', 'c' = nothing
.dw     0x1838                  ;'d' = E
.dw     0x19A9                  ;'e' = D#
.dw     0x16DC                  ;'f' = F
.dw     0x145E                  ;'g' = G
.dw     0x1225                  ;'h' = A
.dw     0xFFFF                  ;'i' = nothing
.dw     0x102A                  ;'j' = B
.dw     0x0F42                  ;'k' = C hi
.dw     0x0D98                  ;'l' = D hi
.dw     0xFFFF, 0xFFFF          ;'m', 'n' = nothing
.dw     0x0E67                  ;'o' = C# hi
.dw     0x0CC8                  ;'p' = D# hi
.dw     0xFFFF, 0xFFFF          ;'q', 'r' = nothing
.dw     0x1B30                  ;'s' = D
.dw     0x1594                  ;'t' = F#
.dw     0x1120                  ;'u' = A#
.dw     0xFFFF                  ;'v' = nothing
.dw     0x1CCE                  ;'w' = C#
.dw     0xFFFF                  ;'x' = nothing
.dw     0x1339                  ;'y' = G#
.dw     0xFFFF                  ;'z' = nothing
.dw     0xFFFF                  ; 26 = nothing
```

```
.org 43
; Seven-segment Lookup Table

.db     0b01110001              ;C
.db     0b10000000, 0b10000000  ; dash
.db     0b11110001              ;E
.db     0b10111110              ;d#
```

```
.db    0b11100001                      ;F
.db    0b01110101                      ;G
.db    0b11100111                      ;A
.db    0b10000000                      ; dash
.db    0b11110100                      ;b
.db    0b01110001                      ;C
.db    0b10110110                      ;d
.db    0b10000000, 0b10000000          ; dash
.db    0b01111001                      ;C#
.db    0b10111110                      ;d#
.db    0b10000000, 0b10000000          ; dash
.db    0b10110110                      ;d
.db    0b11101001                      ;F#
.db    0b11101111                      ;A#
.db    0b10000000                      ; dash
.db    0b01111001                      ;C#
.db    0b10000000                      ; dash
.db    0b01111101                      ;G#
.db    0b10000000, 0b10000000          ; dash
```

```
;===========
EndNote:
       clr     temp
       out     TCCR1A, temp            ;
       reti                            ;


;============
Change:
       in      ZL, UDR                 ; reads data
       subi    ZL, 0x61                ; subtracts 0x61
       cpi     ZL, 26                  ; if ZL is more than 25
       brlo    PC+2                    ;     makes ZL = 26
       ldi     ZL, 26                  ;

       lsl     ZL                      ; multiples ZL by 2
       subi    ZL, -27                 ; adds 27, points to higher byte
       lpm                             ; reads higher byte
       out     OCR1AH, R0              ; stores in OCR1AH
       dec     ZL                      ; points to lower byte
       lpm                             ; reads lower byte
       out     OCR1AL, R0              ; stores in OCR1AL

       subi    ZL, -60                 ; points to second lookup table
       lpm                             ; reads table
```

```
        out     PortB, R0               ; displays result
        mov     temp, R0                ; copies R0 to temp
        andi    temp, 0b00001000        ; masks all but bit 3
        out     PortD, temp             ; copies to PortD to set # LED

        ldi     temp, 0b01000000        ; OC1 toggles with each Output
        out     TCCR1A, temp            ;    Compare interrupt
        clr     temp                    ; resets TCNT0
        out     TCNT0                   ;
        reti                            ;

;========================
Init:   ser     temp                    ; 7 seg code
        out     DDRB, temp              ; PB6: N/C, PB7: Record
        ldi     temp, 0b11111110        ; PD0: RXD
        out     DDRD, temp              ; PD1: TXD

        clr     temp                    ; no pull-ups on PortB
        out     PortB, temp
        out     PortD, temp             ;

        ldi     temp, 0b00000101        ; TC0 is CK/1024
        out     TCCR0, temp             ;
        ldi     temp, 0b01000000        ; no PWM
        out     TCCR1A, temp            ;
        ldi     temp, 0b00001001        ; TC1 is CK, clear TC1
        out     TCCR1B, temp            ;    after compare match

        ldi     temp, 0b01000010        ; enables TC0 interrupt
        out     TIMSK, temp             ; enables TC1 CompA int.
        ldi     temp, 0b00000000        ; disables other interrupts
        out     GIMSK, temp             ;

        ldi     temp, RAMEND
        out     SPL, temp

        ldi     temp, 15                ; baud rate = 9600
        out     UBRR, temp              ;
        ldi     temp, 0b10010000        ; enables RX mode and RX interrupt
        out     UCR, temp               ;

        ldi     NoteH, 0x1E             ; plays a C when first turned on
        ldi     NoteL, 0x84             ;
        out     OCR1AH, NoteH           ;
```

```
      out   OCR1AL, NoteL      ;
      sei                      ; enables interrupts
      clr   ZH                 ; makes sure higher byte of Z is 0
;===================
; Main body of program:

Start:
      rjmp Start
```

PROGRAM P – Computer Controlled Robot
```
;************************************
; written by: John Morton              *
; date:   25/2/02                      *
; version: 1.0                         *
; file saved as: reaction.asm          *
; for AVR: 1200                        *
; clock frequency: 4MHz                *
;************************************

; Program Function: Simple robot which sends and receives commands
; from a computer

.device    at90s2313
.nolist
.include   "C:\Program Files\Atmel\AVR Studio\Appnotes\2313def.inc"
.list


;==========
; Declarations:

.def   temp            =r16
.def   toggle          =r17

.def   data            =r18
.def   speed10         =r19
.def   speed1          =r20

.def   Hundreds        =r21
.def   Tens            =r22
.def   Thousands       =r23
.def   Ones            =r24

.def   DisplayNumber   =r25
```

```
;==============
; Start of Program

        rjmp    Init            ; first line executed
        reti                    ; 001
        reti                    ; 002
        reti                    ; 003
        reti                    ; 004
        reti                    ; 005
        rjmp    Display         ; 006 - T/C0 overflow
        rjmp    Received        ; 007 - UART Rx interrupt
        nop                     ; 008 - UART Empty interrupt
        reti                    ; 09 - UART Tx interrupt
        reti                    ; 0A


;========================
;ASCII to 7 Seg Lookup
.org    16
.db     0b00111111, 0b00000110  ; 0, 1
.db     0b01011011, 0b01001111  ; 2, 3
.db     0b01100110, 0b01101101  ; 4, 5
.db     0b01111101, 0b00000111  ; 6, 7
.db     0b01111111, 0b01101111  ; 8, 9
.org    24
.db     0b01000000, 0b01110111  ; -,  A
.db     0b01111100, 0b00111001  ; b, C
.db     0b01011110, 0b01111001  ; d, E
.db     0b01110001, 0b00111101  ; F, G
.db     0b01110110, 0b00000110  ; H, I
.db     0b00011110, 0b01000000  ; J, -
.db     0b00111000, 0b01000000  ; L, -
.db     0b00110111, 0b00111111  ; N, O
.db     0b01110011, 0b01000000  ; P, -
.db     0b01010000, 0b01101101  ; r, S
.db     0b01111000, 0b00111110  ; t, U
.db     0b01000000, 0b01000000  ; -, -
.db     0b01000000, 0b01101110  ; -, y
.db     0b01011011               ; Z
.org    40
.db     0b01000000, 0b01110111  ; -, A
.db     0b01111100, 0b01011000  ; b, c
.db     0b01011110, 0b01111001  ; d, E
.db     0b01110001, 0b01101111  ; F, g
.db     0b01110100, 0b00000100  ; h, i
```

```
.db    0b00011110, 0b01000000    ; J, -
.db    0b00000110, 0b01000000    ; l, -
.db    0b01010100, 0b01011100    ; n, o
.db    0b01110011, 0b01000000    ; P, -
.db    0b01010000, 0b01101101    ; r, S
.db    0b01111000, 0b00011100    ; t, u
.db    0b01000000, 0b01000000    ; -, -
.db    0b01000000, 0b01101110    ; -, y
.db    0b01011011                ; Z


;===========================
; Command received
Received:
        in    Data, UDR           ; stores received data
        cpi   Data, 0x5D          ; compares data with ']'
        brne  PC+2                ; skips next instruction if not
        rjmp  EndMessage          ; clears T bit

        brtc  PC+2                ; tests T bit (indicates message)
        rjmp  Message

        cpi   Data, 0x67          ; compares data with 'g'
        breq  GoStop              ;

        cpi   Data, 0x74          ; compares data with 't'
        breq  Turning             ;

        cpi   Data, 0x73          ; compares data with 's'
        brne  PC+2                ;
        rjmp  ChangeSpeed         ;

        cpi   Data, 0x2B          ; compares data with '+'
        brne  PC+2                ;
        rjmp  SpeedUp             ;

        cpi   Data, 0x2D          ; compares data with '-'
        brne  PC+2                ;
        rjmp  SlowDown            ;

        cpi   Data, 0x5B          ; compares data with '['
        brne  PC+2                ;
        set                       ; sets T bit

        reti                      ; returns
```

```
GoStop:
      in     temp, TCCR1A      ; reads in current PWM state
      sbrc   temp, 0           ;
      rjmp   Stop              ;
      sbr    temp, 1           ; starts PWM
      out    TCCR1A, temp      ;
      cbi    UCR, RXEN         ; disables receiver
      sbi    UCR, TXEN         ; enables transmitter
      ldi    temp, 0x47        ; "G"
      rcall  Send              ;
      ldi    temp, 0x4F        ; "O"
      rcall  Send              ;
      ldi    temp, 0x21        ; "!"
      rcall  Send              ;
      rjmp   EndMessage

Stop: cbr    temp, 1           ; stops PWM
      out    TCCR1A, temp      ;
      cbi    UCR, RXEN         ; disables receiver
      sbi    UCR, TXEN         ; enables transmitter
      ldi    temp, 0x53        ; "S"
      rcall  Send              ;
      ldi    temp, 0x54        ; "T"
      rcall  Send              ;
      ldi    temp, 0x4F        ; "O"
      rcall  Send              ;
      ldi    temp, 0x50        ; "P"
      rcall  Send              ;
      ldi    temp, 0x21        ; "!"
      rcall  Send              ;
      rjmp   EndMessage

Turning:
in     temp, PortB            ; toggles state of left motor
      eor    temp, toggle      ;
      out    PortB, temp       ;
      cbi    UCR, RXEN         ; disables receiver
      sbi    UCR, TXEN         ; enables transmitter
      ldi    temp, 0x54        ; "T"
      rcall  Send              ;
      ldi    temp, 0x75        ; "u"
      rcall  Send              ;
      ldi    temp, 0x72        ; "r"
      rcall  Send              ;
```

```
        ldi     temp, 0x6E       ; "n"
        rcall   Send             ;
        ldi     temp, 0x69       ; "i"
        rcall   Send             ;
        ldi     temp, 0x6E       ; "n"
        rcall   Send             ;
        ldi     temp, 0x67       ; "g"
        rcall   Send             ;
        rjmp    EndMessage

ChangeSpeed:
        sbis    USR, RXC         ; waits for next byte
        rjmp    ChangeSpeed      ;
        in      speed10, UDR     ; reads tens digit
        mov     data, speed10
        clr     temp
        subi    data, 0x30       ; zeros to 0
Times10:
        breq    CS2              ;
        subi    temp, -10        ;
        dec     data             ;
        rjmp    Times10          ;

CS2:    sbis    USR, RXC         ;
        rjmp    CS2              ;
        in      speed1, UDR      ; reads ones digit
        mov     data, speed1
        subi    data, 0x30       ;
        add     temp, data       ; adds to tens digit

        mov     data, temp       ; multiplies temp by 3
        add     temp, temp       ;
        add     temp, data       ;
        brcc    PC+2             ;
        ldi     temp, 0xFF       ; caps at FF if too high

        out     OCR1AL, temp     ; outputs result

        cbi     UCR, RXEN        ; disables receiver
        sbi     UCR, TXEN        ; enables transmitter
        ldi     temp, 0x53       ; "S"
        rcall   Send             ;
        ldi     temp, 0x70       ; "p"
        rcall   Send             ;
```

```
        ldi    temp, 0x65          ; "e"
        rcall  Send                ;
        ldi    temp, 0x65          ; "e"
        rcall  Send                ;
        ldi    temp, 0x64          ; "d"
        rcall  Send                ;
        ldi    temp, 0x20          ; " "
        rcall  Send                ;
        ldi    temp, 0x73          ; "s"
        rcall  Send                ;
        ldi    temp, 0x65          ; "e"
        rcall  Send                ;
        ldi    temp, 0x74          ; "t"
        rcall  Send                ;
        ldi    temp, 0x20          ; " "
        rcall  Send                ;
        ldi    temp, 0x74          ; "t"
        rcall  Send                ;
        ldi    temp, 0x6F          ; "o"
        rcall  Send                ;
        ldi    temp, 0x20          ; " "
        rcall  Send                ;
        mov    temp, speed10       ; first digit
        rcall  Send                ;
        mov    temp, speed1        ; second digit
        rcall  Send                ;
        rjmp   EndMessage

SpeedUp
:in     temp, OCR1AL              ; reads in current value
        ldi    data, 10
        add    temp, data          ; adds 10
        brcc   PC+2                ; overflowed?
        ldi    temp, 0xFF          ; if so, makes it FF
        out    OCR1AL, temp        ; puts it back
        cbi    UCR, RXEN           ; disables receiver
        sbi    UCR, TXEN           ; enables transmitter
        ldi    temp, 0x53          ; "S"
        rcall  Send                ;
        ldi    temp, 0x70          ; "p"
        rcall  Send                ;
        ldi    temp, 0x65          ; "e"
        rcall  Send                ;
        ldi    temp, 0x65          ; "e"
```

```
        rcall   Send            ;
        ldi     temp, 0x64      ; "d"
        rcall   Send            ;
        ldi     temp, 0x69      ; "i"
        rcall   Send            ;
        ldi     temp, 0x6E      ; "n"
        rcall   Send            ;
        ldi     temp, 0x67      ; "g"
        rcall   Send            ;
        ldi     temp, 0x20      ; " "
        rcall   Send            ;
        ldi     temp, 0x55      ; "U"
        rcall   Send            ;
        ldi     temp, 0x70      ; "p"
        rcall   Send            ;
        rjmp    EndMessage

SlowDown:
        in      temp, OCR1AL    ; reads in current value
        subi    temp, 10        ; subtracts 10
        brcc    PC+2            ; underflowed?
        clr     temp            ; if so, resets to 0
        out     OCR1AL, temp    ; puts it back
        cbi     UCR, RXEN       ; disables receiver
        sbi     UCR, TXEN       ; enables transmitter
        ldi     temp, 0x53      ; "S"
        rcall   Send            ;
        ldi     temp, 0x6C      ; "l"
        rcall   Send            ;
        ldi     temp, 0x6F      ; "o"
        rcall   Send            ;
        ldi     temp, 0x77      ; "w"
        rcall   Send            ;
        ldi     temp, 0x69      ; "i"
        rcall   Send            ;
        ldi     temp, 0x6E      ; "n"
        rcall   Send            ;
        ldi     temp, 0x67      ; "g"
        rcall   Send            ;
        ldi     temp, 0x20      ; " "
        rcall   Send            ;
        ldi     temp, 0x44      ; "D"
        rcall   Send            ;
        ldi     temp, 0x6F      ; "o"
```

```
        rcall  Send              ;
        ldi    temp, 0x77        ; "w"
        rcall  Send              ;
        ldi    temp, 0x6E        ; "n"
        rcall  Send              ;
        rjmp   EndMessage

Message:
        in     ZL, UDR                   ; reads in data
        cbi    UCR, RXEN                 ; disables receiver
        sbi    UCR, TXEN                 ; enables transmitter
        mov    temp, ZL                  ; copies back to PC
        rcall  Send                      ;
        cbi    UCR, TXEN                 ; disables receiver
        sbi    UCR, RXEN                 ; enables transmitter
        subi   ZL, 0x10                  ; subtracts 16
        lpm
        mov    Thousands, Hundreds       ;
        mov    Hundreds, Tens            ;
        mov    Tens, Ones                ;
        mov    Ones, R0                  ;
        reti

EndMessage:
        clt                              ; clears T bit
        cbi    UCR, RXEN                 ; disables receiver
        sbi    UCR, TXEN                 ; enables transmitter
        ldi    temp, 0x0A                ; new line
        rcall  Send                      ;
        ldi    temp, 0x0D                ; carriage return
        rcall  Send                      ;
        cbi    UCR, TXEN                 ; disables receiver
        sbi    UCR, RXEN                 ; enables transmitter
        reti

;==========================================
Send: out    UDR, temp         ;
      sbis   USR, TXC          ;
      rjmp   Send+1            ;
      sbi    USR, TXC
      ret
```

```
;===============================================================
; Display Subroutine

Display:
        inc     DisplayNumber       ;
        cpi     DisplayNumber,4     ;
        brne    PC+2                ;
        clr     DisplayNumber       ;

        ldi     ZL, 21              ; zeros ZL to R21
        add     ZL, DisplayNumber   ;
        ld      temp, Z             ;
        out     PortB, temp         ; outputs temp to Port B

        in      temp, PortD         ;
        lsl     temp                ;
        sbrc    temp, 7             ; gone too far?
        ldi     temp, 0b00001000    ;
        out     PortD, temp         ;

        reti                        ;

;===============================================================

Init:   ldi     temp, 0b11111111    ; PB0-7: outputs
        out     DDRB, temp          ;
        ldi     temp, 0b11111110    ; PD0: input, PD1-6: outputs
        out     DDRD, temp          ;

        ldi     temp, 0b00000000    ; all displays off
        out     PortB, temp         ;
        ldi     temp, 0b00000100    ; selects first display
        out     PortD, temp         ;

        ldi     temp, 0b00000011    ; T/C0 counts at CK/64
        out     TCCR0, temp         ;
        ldi     temp, 0b10000000    ; 8-bit PWM mode on
        out     TCCR1A, temp        ; clears when upcounting
        ldi     temp, 0b00000001    ; T/C1 counts at CK
        out     TCCR1B, temp        ;
        ldi     temp, 0b00000010    ; enables T/C0 overflow
        out     TIMSK, temp         ;

        ldi     temp, 0b10010000    ; turns RXC and TXC interrupts
```

```
    out   UCR, temp                ; enables RX
    ldi   temp, 15                 ;
    out   UBRR, temp               ;

    ldi   temp, RAMEND             ; sets up stack pointer
    out   SPL, temp

    ldi   toggle, 0b10000000       ;
    clr   DisplayNumber            ;
    clr   Thousands                ;
    clr   Hundreds                 ;
    clr   Tens                     ;
    clr   Ones                     ;
    clr   ZH
    sei
    clt                            ; clears T bit
```

```
;===================
; Main body of program:

Start:rjmp  Start                  ;
```

Answers to exercises

Answers to Chapter 1

Answer 1.1: (a)

Largest power of two less than $199 = 128 = 2^7$. Bit $7 = \mathbf{1}$
This leaves $199 - 128 = 71$. 64 is less than 71 so bit $6 = \mathbf{1}$
This leaves $71 - 64 = 7$.   32 is greater than 7 so bit $5 = \mathbf{0}$
16 is greater than 7 so bit $4 = \mathbf{0}$
8 is greater than 7 so bit $3 = \mathbf{0}$
4 is less than 7 so bit $2 = \mathbf{1}$
This leaves $7 - 4 = 3$.   2 is less than 3 so bit $1 = \mathbf{1}$
This leaves $3 - 2 = 1$.   1 equals 1 so bit $0 = \mathbf{1}$

The resulting binary number is: <u>**11000111**</u>

OR...

(b)

Divide 199 by two.	Leaves 99, remainder **1**
Divide 99 by two.	Leaves 49, remainder **1**
Divide 49 by two.	Leaves 24, remainder **1**
Divide 24 by two.	Leaves 12, remainder **0**
Divide 12 by two.	Leaves 6, remainder **0**
Divide 6 by two.	Leaves 3, remainder **0**
Divide 3 by two.	Leaves 1, remainder **1**
Divide 1 by two.	Leaves 0, remainder **1**

So <u>**11000111**</u> is the binary equivalent.

Answer 1.2: (a)

Largest power of two less than $170 = 128 = 2^7$. Bit $7 = \mathbf{1}$
This leaves $170 - 128 = 42$. 64 is greater than 42 so bit $6 = \mathbf{0}$
32 is less than 42 so bit $5 = \mathbf{1}$
This leaves $42 - 32 = 10$.   16 is greater than 10 so bit $4 = \mathbf{0}$
8 is less than 10 so bit $3 = \mathbf{1}$
This leaves $10 - 8 = 2$.   4 is greater than 2 so bit $2 = \mathbf{0}$
2 equals 2 so bit $1 = \mathbf{1}$
Nothing left, so bit $0 = \mathbf{0}$

The resulting binary number is: <u>**10101010**</u>

OR...

(b)

Divide 170 by two.	Leaves 85, remainder **0**
Divide 85 by two.	Leaves 42, remainder **1**
Divide 42 by two.	Leaves 21, remainder **0**
Divide 21 by two.	Leaves 10, remainder **1**
Divide 10 by two.	Leaves 5, remainder **0**
Divide 5 by two.	Leaves 2, remainder **1**
Divide 2 by two.	Leaves 1, remainder **0**
Divide 1 by two.	Leaves 0, remainder **1**

So **10101010** is the binary equivalent.

Answer 1.3:   There are twelve 16s in 199, leaving $199 - 192 = 7$.
So bit $1 = 12 = C$, and bit $0 = 7$. The number is therefore: **C7**.

Answer 1.4:   There are ten 16s in 170, leaving $170 - 160 = 10$. So bit $1 = A$,
and bit $0 = 10 = A$. The number is therefore : **AA**.

Answer 1.5: $1110 = 14 = E$. $0111 = 7$.
The number is therefore **E7**.

Answer 1.6:
$$
\begin{array}{r}
1111 \\
01011010 = 90 \\
+\quad 00001111 = 15 \\
\hline
01101001 = 105
\end{array}
$$

Answer 1.7: $40 = 00101000 \qquad 50 = 00110010$
$-40 = 11010111 + 1 = 11011000$

$$
\begin{array}{r}
1111 \\
11011000 = -40 \\
+\quad 00110010 = 50 \\
\hline
00001010 = 10
\end{array}
$$

Answer 1.8:   8 KB of program memory
512 bytes of EEPROM
512 bytes of SRAM

Answer 1.9:   1. 15 push buttons require five + three = **eight** pins (five input,
three output)
2. Four seven segment displays require four + seven = **eleven**
outputs

Creating a total of **nineteen I/O pins**, hence the smallest AVR in Appendix A is the **4433**.

Answer 1.10:

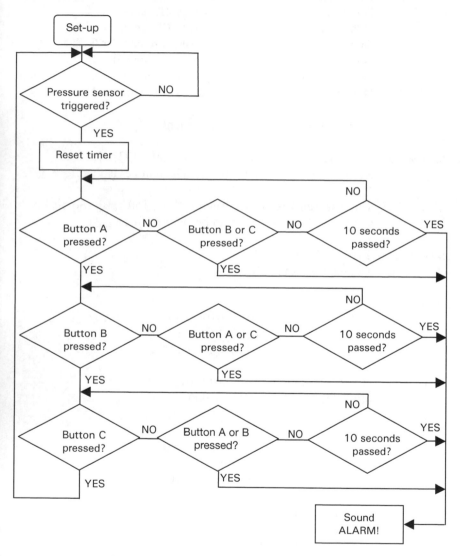

Answer 1.11:

0b00000001	001	0x01
0b00000010	002	0x02
0b00000100	004	0x04

0b00001000	008	0x08
0b00010000	016	0x10
0b00100000	032	0x20
0b01000000	064	0x40
0b10000000	128	0x80
0b00000001	001	0x01 and so on…

Answer 1.12:

In no particular order:

0b00000011	003	0x03
0b00000101	005	0x05
0b00000110	006	0x06
0b00000111	007	0x07

Answer 1.13:

```
ldi     temp, 0b11111110    ; PB0:input, PB1-3: output
out     DDRB, temp          ;   and PB4-7: N/C
ldi     temp, 0b000111      ; PD0-2: outputs, PD3-5: input
out     DDRD, temp          ;   and PD6,7: N/C
ldi     temp, 0b00000001    ; PB0: pull-up, PB1-3: low
out     PortB, temp
clr     temp                ; PD0-2: low, no pull-ups
out     PortD, temp
```

Answers to Chapter 2

Answer 2.1:

```
cbi     PortB, 0            ; turns on LED
rjmp    Start               ; loops back to beginning
```

Answer 2.2:

LEDoff:

```
sbi     PortB, 0            ; turns off LED
rjmp    Start               ; loops back to beginning
```

Answer 2.3:

0:	0b11111100	
1:	0b01100000	or 0b00001100
2:	0b11011010	
3:	0b11110010	
4:	0b01100110	
5:	0b10110110	
6:	0b10111110	
7:	0b11100000	

8:	0b11111110		
9:	0b11110110	or	0b11100110
A:	0b11101110		
b:	0b00111110		
c:	0b00011010		
d:	0b01111010		
E:	0b10011110		
F:	0b10001110		

Answer 2.4:

```
              clr    ZL               ; clears ZL
              clr    ZH               ; clears ZH
ClearLoop:    st     ZL, Z            ; writes ZL to Rx
              inc    ZL               ; moves on to next address
              cpi    ZL, 16           ; gone too far?
              brne   ClearLoop        ; no, so loops back
```

Answer 2.5:

```
Start:        sbic   PinD, 0          ; button pressed?
              rjmp   Start            ; no, so keeps looping
              inc    Counter          ; yes, so adds 1 to Counter
```

Answer 2.6:
```
              cpi    Counter, 10      ; is Counter = 10?
              brne   PC+2             ; no, so skips
              clr    Counter          ; yes, so resets Counter
```

Answer 2.7:
```
              ldi    ZL, 20           ; zeros ZL to R20
              add    ZL, Counter      ; adds Counter to ZL
              ld     temp, Z          ; reads Rx into temp
              out    PortB, temp      ; outputs temp to Port B
              rjmp   Start            ; loops back to Start
```

Answer 2.8:

```
ReleaseWait:  sbis   PinD, 0          ; button released?
              rjmp   ReleaseWait      ; no, so keeps looping
              rjmp   Start            ; yes, so loops back to start
```

Answer 2.9:   Rising edge, external count. So the number is: **0b00000111**.

Answer 2.10:
```
UpTest:       sbic   PinD, 1          ; checks speed-up button
              rjmp   Timer            ; not pressed, jumps
              dec    Speed            ; speeds up time
              brne   ReleaseUp        ; jumps to ReleaseUp if not 0
```

	inc	Speed	; adds one to Speed
ReleaseUp:	sbis	PinD, 1	; waits for button to be released
	rjmp	ReleaseUp	;

Answer 2.11: mov   Counter, Speed   ; copies Speed into Counter

Answer 2.12: Moves 03C into PC.

Answer 2.13: **400 000** clock cycles. Divide by 5 = **80 000** = **0x13880**
Split up over three registers, so their initial values will be:
0x80, **0x38**, and **0x01**.

Answer 2.14:

Debounce:	ldi	Delay1, 0x80	; sets up counting registers
	ldi	Delay2, 0x38	;
	ldi	Delay3, 0x01	;
Loop:	subi	Delay1, 1	; inserts delay
	sbci	Delay2, 0	;
	sbci	Delay3, 0	;
	brcc	Loop	;
	ret		; returns from subroutine

Answer 2.15:

Start:	ldi	temp, 0b00010001	; motorists: green
	out	PortB, temp	; pedestrians: temp

Answer 2.16: sbic   PinD, 0   ; tests button
                rjmp   Start   ; not pressed

Answer 2.17: sbi   PortB, 5   ; turns on WAIT light

Answer 2.18:

Loop:	rcall	Timer	; keeps timing
	brts	Loop	; stays in loop until T is clear

Answer 2.19: sbi   PortB, 1   ; motor amber on
                cbi   PortB, 0   ; motor green off

Answer 2.20: ldi   temp, 0b00001100 ; motorists: red
                out   PortB, temp   ; pedestrians: green

Answer 2.21: ldi   temp, 16   ; 8 seconds delay
EightSeconds:
                rcall HalfSecond   ;

```
                dec      temp              ;
                brne     EightSeconds      ;
```

Answer 2.22:
```
        ldi      temp, 0b00001010   ; motorists: amber
        out      PortB, temp        ; pedestrians: green
```

Answer 2.23: 0b10110011 → **0b01001100**

Answer 2.24:
```
        ldi      tog, 0b00001010    ; sets up tog register
        in       temp, PinB         ; reads in state of lights
        eor      temp, tog          ; toggles
        out      PortB, temp        ; outputs
```

Answer 2.25:
```
        ldi      tog, 0b00001010    ; sets up tog register
        ldi      Counter, 8         ; sets up Counter register
```

```
FlashLoop:  rcall    HalfSecond         ; waits ½ a second
            in       temp, PinB         ; reads in state of lights
            eor      temp, tog          ; toggles
            out      PortB, temp        ; outputs
            dec      Counter            ; does this 8 times
            brne     FlashLoop          ;
```

Answer 2.26:
```
        set                         ; sets T bit
        rjmp            Start        ; loops back to Start
```

Answer 2.27: **1 228 800** clock cycles. Divide by 5 = **245 760** = **0x3C000**
Split up over three registers, so their initial values will be:
0x00, 0xC0, and **0x03.**

```
HalfSecond: clr      Delay1             ; sets up counting registers
            ldi      Delay2, 0xC0       ;
            ldi      Delay3, 0x03       ;
```

```
HalfLoop:   subi     Delay1, 1          ; inserts delay
            sbci     Delay2, 0          ;
            sbci     Delay3, 0          ;
            brcc            HalfLoop    ;
            ret                         ;
```

Answer 2.28:
```
Timer:      brts     PC+2               ; test T bit, skip if set
            ret                         ; returns if T is clear
```
Answer 2.29:
```
            in       temp, TCNT0        ; reads Timer 0 into temp
```

```
cpse   temp, Mark240      ; compares temp with Mark240
ret                       ; if not equal returns
subi   Mark240, -240      ; adds 240 to Mark240
dec    Count250           ; subtracts one from Count250
breq            PC+2      ; if zero, skips next line
ret                       ; if not zero returns

ldi    Count250, 250      ; resets Count250
clt                       ; clears T bit
ret                       ;
```

Answer 2.30:

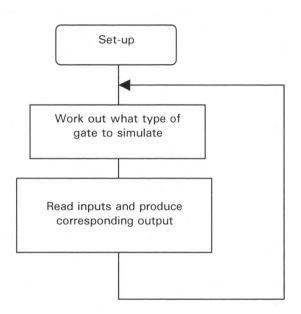

Answer 2.31: **lsr**

Answer 2.32:

```
Start:      in     ZL, PinB          ; reads in PinB
            andi   ZL, 0b001110      ; masks 0, 4 and 5
            lsr    ZL                ; rotates
```

Answer 2.33:
```
subi   ZL, -2             ; adds 2 to ZL
lpm                       ; reads lookup table into R0
```

Answer 2.34:	NAND	1110 ->	00**110010**
	NOR	1000 ->	00**100000**
	ENOR	1001 ->	00**100001**
	EOR	0110 ->	00**010010**
	NOT	1100 ->	00**110000**
	Buffer	0011 ->	00**000011**

Answer 2.35:

```
.dw        0b0011001000100000        ; NAND and NOR
.dw        0b0010000100010010        ; ENOR and EOR
.dw        0b0011000000000011        ; NOT and buffer
```

Answer 2.36:

```
mov     temp, R0          ; copies R0 to temp
ori     temp, 0b11110     ; forces bits 1-4 high
out     PortB, temp       ; outputs result
rjmp    Start             ; loops back to Start
```

Answer 2.37:

64 ms = 256 000 cycles
Divide by 8 = 32 000 decrements = 0x7D00
Delay1 initialized to **0x00** and **Delay2** initialized to **0x7D**

Answer 2.38:

```
DoneHi:   in      lowerbyte, TCNT0   ; immediately stores TCNT0
          cp      lowerbyte, temp    ; compares with previous value
          brsh    Divide64           ; jumps to Divide64 if OK
          inc     upperbyte          ; increments higher byte
          cpi     upperbyte, 0xFA    ; has it gone too far?
          breq    TooHigh            ; skips to TooHigh if so
```

Answer 2.39:
```
          lsr     upperbyte          ; rotates right, bit 7 = 0
          ror     lowerbyte          ; rotates right, bit 7 = carry
                                     ;     flag
```

Answer 2.40:

```
Divide64: ldi     temp, 6            ; sets up temp with 6
          lsr     upperbyte          ; divides two-byte word by 2
          ror     lowerbyte          ;
          dec     temp               ; does this 6 times
          brne    Divide64+1         ; keeps looping until finished
```

Answer 2.41:
```
          cpi     upperbyte,0        ; higher byte 0?
          brne    PC+3               ; skips next 2 instructions
```

```
        cpi      lowerbyte, 0          ; lower byte 0?
        breq     LowSpeed              ; jumps to if freq < 1kHz
```

Answer 2.42: 2 000 000 clock cycles to waste.
14 cycles in loop => 142 857 = 0x022E09
Therefore set up registers with **0x09**, **0x2E**, and **0x02**.

```
             ldi      Delay1, 0x09         ; sets ups delay registers
             ldi      Delay2, 0x2E         ;
             ldi      Delay3, 0x02         ;
HalfSecond:
             rcall    Display              ; calls display for half a second
             subi     Delay1, 1            ;
             sbci     Delay2, 0            ;
             sbci     Delay3, 0            ;
             brcc     HalfSecond           ;
             rjmp     Start                ; loops back to Start
```

Answer 2.43:
```
TooHigh:     ldi      Hundreds, 11         ; code for a -
             ldi      Tens, 10             ; code for a H
             ldi      Ones, 1              ; code for a I
             rjmp     HalfSecond-3         ; displays for half a second
```

Answer 2.44:
```
Display:     dec      DisplayCounter       ; changes display every 50 visits
             breq     PC+2                 ; skips if 50th time
             ret                           ; returns
             wdr                           ; pats the dog
             ldi      DisplayCounter, 50   ; resets DisplayCounter
```

Answer 2.45:
```
             inc      DisplayNumber        ; increments DisplayNumber
             cpi      DisplayNumber,3      ; has it reached 3?
             brne         PC+2             ; no, so skips
             clr      DisplayNumber        ; yes, so clears
```

Answer 2.46:
```
             ldi      ZL, 26               ; initializes ZL to R26
             add      ZL, DisplayNumber    ; points to right digit
             ld       temp, Z              ; loads value into temp
             clr      ZL                   ; zeros ZL to R0
             add      ZL, temp             ; adds temp to ZL
             ld       temp, Z              ; reads Rx into temp
             sbic     PortB, 7             ; tests kHz LED
             ori      temp, 0b10000000     ; if it's on, keeps it on
             out      PortB, temp          ; outputs temp to Port B
```

Answer 2.47:

```
            in      temp, PinD           ; reads in PinD
            lsl     temp                 ; rotates left
            sbrc    temp, 3              ; tests bit 3 of result
            ldi     temp, 0b00000001     ; resets if gone too far
            out     PortD, temp          ; outputs result to Port D
            ret                          ; returns from the subroutine
```

Answer 2.48:

```
LowSpeed:   ldi     temp, 0b00000001     ; sets TCNT0 to count at CK
            out     TCCR0, temp          ;
            clr     Delay2               ; resets delay registers
            clr     Delay3               ;
            cbi     PortB, 7             ; clears PB7 to turn on Hz LED
```

Answer 2.49:

```
            ldi     Counter, 2           ; sets up Counter to 2
            clr     Delay1               ; resets Delay1 and TCNT0
            out     TCNT0, Delay1        ;
```

Answer 2.50:

```
            in      store, PinD          ; stores initial value
LowLoop:    in      store2, PinD         ; reads in current value
            eor     store2, store        ; compares initial and current
            sbrc    store2, 4            ; skips if PD4 unchanged
            rjmp    Change               ; jumps if PD4 changes
```

Answer 2.51:

```
            rcall   Display              ; keeps displays going

            mov     temp2, Delay1        ; stores old value
            in      Delay1, TCNT0        ; reads in new value
            cp      Delay1, temp2        ; compares old and new
            brsh    LowLoop              ; loops back of new > old

            inc     Delay2               ; increment higher byte
            brne    LowLoop              ; test if zero, loops if isn't
            inc     Delay3               ; increments highest byte
            cpi     Delay3, 0x3E         ; too slow?
            breq    TooSlow              ; yes
            rjmp    LowLoop              ; no, so loops back
```

Answer 2.52:

```
Divide:     sub     temp, Delay1         ; subtracts result from 400 000
            sbc     temp2, Delay2        ;
            sbc     temp3, Delay3        ;
            brcs    DoneDividing         ; if Carry set, finished dividing
```

```
             inc     lowerbyte          ; if not set, adds 1 to answer
             brne    Divide             ; overflow?
             inc     upperbyte          ; yes, so increments higher byte
             rjmp    Divide             ; keeps looping
```

Answer 2.53:
```
DoneDividing:
             rcall   DigitConvert       ; converts answer into digits
             rjmp    LowSpeed           ; loops back to beginning
```

Answer 2.54:
```
TooSlow:     clr     temp               ;
             out     PortD, temp        ; turns off Displays
             sleep                      ; goes to sleep
```

Answers to Chapter 4

Answer 4.1:
```
             rjmp    Init               ; first line executed
             rjmp    ExtInt             ; handles external interrupt
             rjmp    OverflowInt        ; handles TCNT0 interrupt
```

Answer 4.2:
```
             ldi     temp, 0b01000000   ; sets bit 6 – enables External
             out     GIMSK, temp        ;    INT0 interrupt
             clr     temp               ; selects low level interrupt
             out     MCUCR, temp        ;
             ldi     temp, 0b00000010   ; enables TCNT0 interrupt
             out     TIMSK, temp        ;
```

Answer 4.3:
```
Start:       rcall   Display            ; keeps display going
             sbic    PinD, 1            ; waits for Ready button
             rjmp    Start              ; keeps looping until pressed
```

Answer 4.4:
```
             mov     temp, Random       ; multiplies by 5 and...
             add     Random, temp       ;
             add     Random, temp       ;
             add     Random, temp       ;
             add     Random, temp       ;
             inc     Random             ; ...adds 1
```

Answer 4.5:
```
             mov     CountX, Random     ;
             lsr     CountX             ; divides by 2
             subi    CountX, -60        ;  and adds 60
```

Answer 4.6:

```
          clr     TimeH              ; reset timing register
          out     PortB, TimeH       ; turns off display
          ldi     temp, 0b0100000    ; resets INT0 interrupt flag
          out     GIFR               ;
          ldi     temp, 0b00000010   ; resets TC0 OVF interrupt flag
          out     TIFR               ;
```

Answer 4.7:

```
          sei                        ; enables interrupts
Loopy:    brid    Start              ; skips out when interrupts
          rjmp    Loopy              ;    disabled
```

Answer 4.8:

```
ExtInt:   sbis    PinD, 0            ; tests LED
          rjmp    Cheat              ;
          clr     temp               ; stops TCNT0
          out     TCCR0, temp        ;
          in      TimeL, TCNT0       ; reads in TCNT0 value
          in      temp, TIFR         ; test for TCNT0 overflow
          sbrc    temp, 1            ;
          inc     TimeH              ;
          subi    TimeL, 0xA2        ; subtracts back 0xA2 from
          sbci    TimeH, 0           ;    total reaction time
          ldi     temp, 0b00000101   ; restarts TCNT0 at CK/1024
          out     TCCR0, temp        ;
```

Answer 4.9:

```
Cheat:    ldi     Hundreds, 10       ; b
          ldi     Tens, 11           ; A
          ldi     Ones, 12           ; d
          ret
```

Answer 4.10:

```
          clr     TimeL              ; resets result registers
          clr     TimeH              ;
Divide12: subi    temp, 12           ; subtracts 12 from total
          sbci    tempH, 0           ;
          brcs    DoneDividing       ; skips out when there's a carry
          inc     TimeL              ; increment lower byte
          brne    Divide12           ; lower byte = 0?
          inc     TimeH              ; yes, so increment higher byte
          rjmp    Divide12           ; loop back
```

Answer 4.11:

```
Start:    ldi     temp, 0b00001000   ; puts initial value in Port D
          out     PortD, temp        ;
```

```
                sbis    ACSR, 5          ; checks AC result
                cbi     PortD, 3         ; clears bit 3 if it is low
                sbi     PortD, 2         ; sets bit 2 in either case

Answer 4.12:    sbis    ACSR, 5          ; checks AC result
                cbi     PortD, 2         ; clears bit 2 if it is low
                sbi     PortD, 1         ; sets bit 1 in either case

                sbis    ACSR, 5          ; checks AC result
                cbi     PortD, 1         ; clears bit 1 if it is low
                sbi     PortD, 0         ; sets bit 0 in either case

                sbis    ACSR, 5          ; checks AC result
                cbi     PortD, 0         ; clears bit 0 if it is low

Answer 4.13:    in      temp, PortD      ; reads in final answer
                swap    temp             ; swap bits 0-3 for 4-7
                out     PortB, temp      ; outputs result
                rjmp    Start            ;
```

Answer 4.14: 0b11100011 → ADCSR
0b00000000 → ADMUX

Answer 4.15:

```
Start:          cbi     ADMUX, 0         ; selects ADC0 input
                sbi     ADCSR, ADSC      ; starts conversion
                sbic    ADCSR, ADSC      ; has conversion finished?
                rjmp    Start+2          ; no, so keeps waiting

Answer 4.16:    in      Desired, ADCH    ; reads in 8-bits of answers
                com     Desired          ; 5 - answer

                sbi     ADMUX, 0         ; selects ADC1 input
                sbi     ADCSR, ADSC      ; starts conversion on ADC1
Wait:           sbic    ADCSR, ADSC      ; has conversion finished?
                rjmp    Wait             ; no, so keeps waiting

Answer 4.17:    in      Actual, ADCH     ; reads in V of actual output
                cp      Actual, Desired  ; compares actual with desired
                brlo    TooLow           ; too low?
                cp      Desired, Actual  ;
                brlo    TooHigh          ; too high?
                cbi     DDRB, 0          ; actual = desired so makes PB0
                rjmp        Start        ;    an input and loops to Start
```

Answer 4.18:

```
TooHigh:     sbi     DDRB, 0        ; makes PB0 and output
             sbi     PortB, 0       ; makes PB0 5V
             rjmp    Start          ; loops back to Start

TooLow:      sbi     DDRB, 0        ; makes PB0 an output
             cbi     PortB, 0       ; makes PB0 0V
             rjmp    Start          ; loops back to Start
```

Answer 4.19:
```
             clr     Address        ; first address is 0x00
             ldi     Data, 0x30     ; ASCII for "0" is 0x30

ASCIILoop:   out     EEAR, Address  ;
             out     EEDR, Data     ;
             sbi     EECR, 1        ; initiates write
EEWait:      sbic    EECR, 1        ; waits for write to finish
             rjmp    EEWait         ; loops until EECR, 1 is cleared
             inc     Address        ; selects next address
             inc     Data           ; selects next ASCII code
             cpi     Data, 0x3A     ; finished doing numbers?
             brne    PC+2           ; skips if not finished
             ldi     Data, 0x41     ; ASCII for "A" is 0x41
             cpi     Data, 0x47     ; finished completely?
             brne    ASCIILoop      ; yes, finished
```

Answer 4.20:
```
             0b00001101 → TCCR1B    ; T/C1 prescaled at CK/1024
                                    ; reset T/C1 on compare match
             0b01000000 → TIMSK     ; enables output compare int.
             0x0F       → OCR1AH    ; 4MHz / 1024 = 3906Hz
             0x42       → OCR1AL    ; 3906 = 0xF42
```

Answer 4.21:
```
ToggleOut:   in      temp, PortB    ; reads in Port B
             com     temp           ; inverts bits
             out     PortB, temp    ; outputs to Port B
             reti                   ; returns
```

Answer 4.22:
```
ChangeNote:  dec     Length         ;
             breq    PC+2           ; skips on when enough time
             reti                   ;     has passed

Rest:        in      temp, TIFR     ; waits until T/C0 overflow
             sbrs    temp, 1        ;     interrupt flag
```

```
                rjmp    Rest                    ;
                ldi     temp, 0b00000010        ; resets interrupt flag
                out     TIFR, temp              ;
```

Answer 4.23:
Read_EEPROM:
```
                out     EEARL, address          ; select address
                sbi     EECR, 0                 ; initiate read

                in      ZL, EEDR                ; reads EEPROM
                andi    ZL, 0b00001111          ; masks higher nibble
                cpi     ZL, 0x0C                ; compares with 0xC
                breq    Reset                   ; repeats melody if equal
                brlo    PC+2                    ; is ZL < 0xC
                ldi     ZL, 0x00                ; if it is selects a 'C' (0x0)
```

Answer 4.24:
```
                lsl     ZL                      ; multiplies ZL by two
                subi    ZL, -0x26               ; adds 26 to point to table
                lpm                             ; reads table
                mov     NoteL, R0               ; stores lower byte
                inc     ZL                      ; moves to next address
                lpm                             ; reads table
                mov     NoteH, R0               ; stores higher byte
```

Answer 4.25:
```
                in      temp, EEDR              ; reads in the byte
                swap    temp                    ; swaps nibbles
                andi    temp, 0b00000011        ; selects correct bits
GetOctave:      breq    GetLength               ; skips if 0
                lsr     NoteH                   ; rotates higher byte
                ror     NoteL                   ; rotates lower byte with carry
                dec     temp                    ; repeats for each octave
                rjmp    GetOctave               ;
```

Answer 4.26:
```
GetLength:      out     OCR1AH, NoteH           ; stores note values in
                out     OCR1AL, NoteL           ; output compare registers
                in      temp, EEDR              ; reads in EEPROM again
                andi    temp, 0b11000000        ; masks bits
                swap    temp                    ; swaps nibbles
                lsr     temp                    ; rotates once
                subi    temp, -2                ; adds two
                mov     Length, temp            ; moves into Length
                reti                            ; returns
```

Answers to Chapter 5

Answer 5.1:
```
             ldi     Counter, 8        ; sets up counter with 8
             clr     parityreg         ; resets register
Parity:      lsr     temp              ; rotates temp to the right
             brcc    PC+2              ;
             inc     parityreg         ;
             dec     Counter, 8        ; does this 8 times
             brne    Parity            ;
```

Bit 0 of **parityreg** is now a parity bit for **temp**.

Answer 5.2:
```
Change:      in      ZL, UDR           ; reads data
             subi    ZL, 0x61          ; subtracts 0x61
             cpi     ZL, 26            ; if ZL is more than 25
             brlo    PC+2              ;    makes ZL = 26
             ldi     ZL, 26            ;

             lsl     ZL                ; multiples ZL by 2
             subi    ZL, -27           ; adds 27, points to higher byte
             lpm                       ; reads higher byte
             out     OCR1AH, R0        ; stores in OCR1AH
             dec     ZL                ; points to lower byte
             lpm                       ; reads lower byte
             out     OCR1AL, R0        ; stores in OCR1AL
```

Answer 5.3:
```
             subi    ZL, -60           ; points to second lookup table
             lpm                       ; reads table
             out     PortB, R0         ; displays result
             mov     temp, R0          ; copies R0 to temp
             andi    temp, 0b00001000  ; masks all but bit 3
             out     PortD, temp       ; copies to PortD to set # LED
```

Answer 5.4:
```
             ldi     temp, 0b01000000  ; OC1 toggles with each Output
             out     TCCR1A, temp      ;    Compare interrupt
             clr     temp              ; resets TCNT0
             out     TCNT0             ;
             reti
```

Answer 5.5:
```
EndNote:     clr     temp              ; disconnects OC1 pin from
             out     TCCR1A, temp      ;    OC interrupt
             reti                      ;
```

Answer 5.6:

```
.macro      skeq                        ; calls this macro skeq
            breq PC+2                   ;
.endmacro
```

Answer 5.7:

```
.macro      HiWait                      ; calls this macro HiWait
            sbis    @0, @1              ; tests input
            rjmp    PC-1                ; keeps looping until input is set
.endmacro
```

Answer 5.8:

```
Display:    inc     DisplayNumber       ; selects next display
            cpi     DisplayNumber,4     ; gone too far?
            brne PC+2                   ;
            clr     DisplayNumber       ; yes, so resets to first

            ldi     ZL, 21              ; zeros ZL to R21
            add     ZL, DisplayNumber   ; adds display number
            ld      temp, Z             ; reads value
            out     PortB, temp         ; outputs temp to Port B

            in      temp, PortD         ; reads in current value
            lsl     temp                ; moves to next display
            sbrc    temp, 7             ; gone too far?
            ldi     temp, 0b00001000    ; resets to first display
            out     PortD, temp         ; outputs result
            reti                        ; returns enabling interrupts
```